Exploring the Ocean Environment

GIS Investigations for the Earth Sciences

Michelle K. Hall-Wallace
C. Scott Walker
Jennifer A. Weeks
Larry P. Kendall
Science Education Solutions, Inc.

THOMSON

BROOKS/COLE

Australia • Canada • Mexico • Singapore • Spain • United Kingdom • United States

THOMSON

™

BROOKS/COLE

Editor: *Keith Dodson*
Assistant Editor: *Carol Ann Benedict*
Editorial Assistant: *Melissa Newt*
Marketing Manager: *Melanie Banfield*
Advertising Project Manager: *Kelley McAllister*
Project Manager, Editorial Production: *Kelsey McGee*

Print/Media Buyer: *Rebecca Cross*
Permissions Editor: *Chelsea Junget*
Cover Designer: *Denise Davidson*
Cover Image: *©Royalty-Free/CORBIS*
Cover Printer: *Banta Book Group/Menasha*
Printer: *Banta Book Group/Menasha*

Printed in the United States of America

1 2 3 4 5 6 7 08 07 06 05 04

For more information about our products, contact us at:
Thomson Learning Academic Resource Center
1-800-423-0563

For permission to use material from this text or product, submit a request online at http://www.thomsonrights.com. Any additional questions about permissions can be submitted by email to thomsonrights@thomson.com.

All products used herein are used for identification purposes only and may be trademarks or registered trademarks of their respective owners.

Maps and screen shots include data from ESRI Data and Maps. Data Copyright © ESRI 2002.

ESRI and ArcView are registered trademarks in the United States and are either trademarks or registered trademarks in all other countries in which they are used. The ArcView logo is a trademark of Environmental Systems Research Institute, Inc.

This material is based upon work supported by the National Science Foundation under Grants No. DUE-9555205, IMD-9986613, EAR-9809704, and GRD-9979670.

Any opinions, findings, and conclusions or recommendations expressed in this material are those of the authors and do not necessarily reflect the views of the National Science Foundation.

ISBN: 0-534-42350-7

Thomson Brooks/Cole
10 Davis Drive
Belmont CA 94002
USA

Asia
Thomson Learning
5 Shenton Way #01-01
UIC Building
Singapore 068808

Australia
Thomson Learning
102 Dodds Street
Southbank, Victoria 3006
Australia

Canada
Nelson
1120 Birchmount Road
Toronto, Ontario M1K 5G4
Canada

Europe/Middle East/Africa
Thomson Learning
High Holborn House
50/51 Bedford Row
London WC1R 4LR
United Kingdom

Latin America
Thomson Learning
Seneca, 53
Colonia Polanco
11560 Mexico D.F.
Mexico

Spain
Paraninfo
Calle Magallanes, 25
28015 Madrid, Spain

Acknowledgments

The authors wish to thank the many students, teachers, and scientists who, through their use of these materials, provided critical reviews and helped us develop insight into how GIS can be most effectively used as a learning and teaching tool.

A significant number of people contributed directly or indirectly to the development of this module, but a few are especially notable. Particular thanks go to Alan Kelley, Kathy Krucker, Kathy Likos, Graciela Rendon-Coke, Anthony Occhiuzzi, and Anne Thames who tested the investigations in their classrooms and provided valuable feedback that helped us to improve the content and design of the activities. Anne Ortiz, Jessica Walker, and James Kevin contributed significantly to the final rounds of editing and revision, while Robert Butler, Miles Logsdon, David Smith, and Cheryl Greengrove provided content reviews. We also appreciate the considerable efforts of Ted Stude for his assistance in preparing data and illustrations.

We are indebted to the numerous scientists who took the time to learn about our project and share critical research data or expertise that added greatly to the quality of the investigations. Specifically, we wish to thank Dr. Tom Garrison (Orange Coast College and University of Southern California) for his ongoing input and support, as well as Dr. Nancy Rabalais (Louisiana University Marine Consortium) and Dr. Robert Diaz (Virginia Institute of Marine Science) for sharing their data on the Mississippi River and global dead zones, respectively.

We are grateful to the agencies, publishers, and individuals who have given us permission to include their outstanding illustrations, stories, and photos. Finally, we would like to acknowledge the National Science Foundation for their substantial support of this project through grants DUE-9555205, IMD-9986613, EAR-9809704, and GRD-9979670.

The SAGUARO Project
Michelle K. Hall, Director
Science Education Solutions, Inc.
Los Alamos Research Park • 4200 West Jemez Drive
Synergy Center Suite 301 • Los Alamos NM 87544
saguaro@scieds.com

Science And GIS Unlocking Analysis & Research Opportunities
http://www.scieds.com/saguaro

ESRI Software License Agreement

This is a license agreement and not an agreement for sale. This license agreement (Agreement) is between the end user (Licensee) and Environmental Systems Research Institute, Inc. (ESRI), and gives Licensee certain limited rights to use the proprietary ESRI® desktop software and software updates, sample data, online and/or hard-copy documentation and user guides, including updates thereto, and software keycode or hardware key, as applicable (hereinafter referred to as "Software, Data, and Related Materials"). All rights not specifically granted in this Agreement are reserved to ESRI.

Reservation of Ownership and Grant of License: ESRI and its third party licensor(s) retain exclusive rights, title, and ownership of the copy of the Software, Data, and Related Materials licensed under this Agreement and, hereby, grants to Licensee a personal, nonexclusive, nontransferable license to use the Software, Data, and Related Materials based on the terms and conditions of this Agreement. From the date of receipt, Licensee agrees to use reasonable effort to protect the Software, Data, and Related Materials from unauthorized use, reproduction, distribution, or publication.

Copyright: The Software, Data, and Related Materials are owned by ESRI and its third party licensor(s) and are protected by United States copyright laws and applicable international laws, treaties, and/or conventions. Licensee agrees not to export the Software, Data, and Related Materials into a country that does not have copyright laws that will protect ESRI's proprietary rights. Licensee may claim copyright ownership in the Simple Macro Language (SML™) macros, AutoLISP® scripts, AtlasWare™ scripts, and/or Avenue™ scripts developed by Licensee using the respective macro and/or scripting language.

Permitted Uses:
- Licensee may use the number of copies of the Software, Data, and Related Materials for which license fees have been paid on the computer system(s) and/or specific computer network(s) for Licensee's own internal use. Licensee may use the Software, Data, and Related Materials as a map/data server engine in an Internet and/or Intranet distributed computing network or environment provided the appropriate, additional license fees are paid. If the Software, Data, and Related Materials contain dual media (i.e., both 3.5-inch diskettes and CD–ROM), then Licensee may only use one (1) set of the dual media provided. Licensee may not use the other media on another computer system(s) and/or specific computer network(s), or loan, rent, lease, or transfer the other media to another user.
- Licensee may install the number of copies of the Software, Data, and Related Materials for which license or update fees have been paid onto the permanent storage device(s) on the computer system(s) and/or specific computer network(s).
- Licensee may make routine computer backups but only one (1) copy of the Software, Data, and Related Materials for archival purposes during the term of this Agreement unless the right to make additional copies is granted to Licensee in writing by ESRI.
- Licensee may use, copy, alter, modify, merge, reproduce, and/or create derivative works of the online documentation for Licensee's own internal use. The portions of the online documentation merged with other software, hard copy, and/or digital materials shall continue to be subject to the terms and conditions of this Agreement and shall provide the following copyright attribution notice acknowledging ESRI's proprietary rights in the online documentation: "Portions of this document include intellectual property of ESRI and are used herein by permission. Copyright © 200_ Environmental Systems Research Institute, Inc. All Rights Reserved."
- Licensee may use the Data that are provided under license from ESRI and its third party licensor(s) as described in the Distribution Rights section of the online Data Help files.

Uses Not Permitted:
- Licensee shall not sell, rent, lease, sublicense, lend, assign, time-share, or transfer, in whole or in part, or provide unlicensed third parties access to prior or present versions of the Software, Data, and Related Materials, any updates, or Licensee's rights under this Agreement.
- Licensee shall not reverse engineer, decompile, or disassemble the Software, or make any attempt to unlock or bypass the software keycode and/or hardware key used, as applicable, subject to local law.
- Licensee shall not make additional copies of the Software, Data, and/or Related Materials beyond that described in the Permitted Uses section above.
- Licensee shall not remove or obscure any ESRI copyright or trademark notices.
- Licensee shall not use this software for more than one hundred twenty (120) days from the date that the software was installed. At the end of this period, users must remove the time limited software from their computers or purchase fully licensed software. Students and instructors in the United States may purchase fully licensed individual copies of the software from ESRI telesales at 1-800-GIS-XPRT. (1-800-447-9778)

Term: The license granted by this Agreement shall commence upon Licensee's receipt of the Software, Data, and Related Materials and shall continue until such time that (1) Licensee elects to discontinue use of the Software, Data, and Related Materials and terminates this Agreement or (2) ESRI terminates for Licensee's material breach of this Agreement. Upon termination of this Agreement in either instance, Licensee shall return to ESRI the Software, Data, Related Materials, and any whole or partial copies, codes, modifications, and merged portions in any form. The parties hereby agree that all provisions, which operate to protect the rights of ESRI, shall remain in force should breach occur.

Limited Warranty: ESRI warrants that the media upon which the Software, Data, and Related Materials are provided will be free from defects in materials and workmanship under normal use and service for a period of sixty (60) days from the date of receipt. The Data herein have been obtained from sources believed to be reliable, but its accuracy and completeness, and the opinions based thereon, are not guaranteed. Every effort has been made to provide accurate Data in this package. The Licensee acknowledges that the Data may contain some nonconformities, defects, errors, and/or omissions. ESRI and third party licensor(s) do not warrant that the Data will meet Licensee's needs or expectations, that the use of the Data will be uninterrupted, or that all nonconformities can or will be corrected. ESRI and the respective third party licensor(s) are not inviting reliance on these Data, and Licensee should always verify actual map data and information. The Data contained in this package are subject to change without notice.

EXCEPT FOR THE ABOVE EXPRESS LIMITED WARRANTIES, THE SOFTWARE, DATA, AND RELATED MATERIALS CONTAINED THEREIN ARE PROVIDED "AS IS," WITHOUT WARRANTY OF ANY KIND, EITHER EXPRESS OR IMPLIED, INCLUDING, BUT NOT LIMITED TO, THE IMPLIED WARRANTIES OF MERCHANTABILITY AND FITNESS FOR A PARTICULAR PURPOSE.

Exclusive Remedy and Limitation of Liability: During the warranty period, ESRI's entire liability and Licensee's exclusive remedy shall be the return of the license fee paid for the Software, Data, and Related Materials in accordance with the ESRI Customer Assurance Program for the Software, Data, and Related Materials that do not meet ESRI's Limited Warranty and that are returned to ESRI or its dealers with a copy of Licensee's proof of payment.

ESRI shall not be liable for indirect, special, incidental, or consequential damages related to Licensee's use of the Software, Data, and Related Materials, even if ESRI is advised of the possibility of such damage.

Waivers: No failure or delay by ESRI in enforcing any right or remedy under this Agreement shall be construed as a waiver of any future or other exercise of such right or remedy by ESRI.

Order of Precedence: Any conflict and/or inconsistency between the terms of this Agreement and any FAR, DFAR, purchase order, or other terms shall be resolved in favor of the terms expressed in this Agreement, subject to the U.S. Government's minimum rights unless agreed otherwise.

Export Regulations: Licensee acknowledges that this Agreement and the performance thereof are subject to compliance with any and all applicable United States laws, regulations, or orders relating to the export of computer software or know-how relating thereto. ESRI Software, Data, and Related Materials have been determined to be Technical Data under United States export laws. Licensee agrees to comply with all laws, regulations, and orders of the United States in regard to any export of such Technical Data. Licensee agrees not to disclose or reexport any Technical Data received under this Agreement in or to any countries for which the United States Government requires an export license or other supporting documentation at the time of export or transfer, unless Licensee has obtained prior written authorization from ESRI and the U.S. Office of Export Control. The countries restricted at the time of this Agreement are Cuba, Iran, Iraq, Libya, North Korea, Serbia, and Sudan.

U.S. Government Restricted/Limited Rights: Any software, documentation, and/or data delivered hereunder is subject to the terms of the License Agreement. In no event shall the Government acquire greater than RESTRICTED/LIMITED RIGHTS. At a minimum, use, duplication, or disclosure by the Government is subject to restrictions as set forth in FAR §52.227-14 Alternates I, II, and III (JUN 1987); FAR §52.227-19 (JUN 1987) and/or FAR §12.211/12.212 (Commercial Technical Data/Computer Software); and DFARS §252.227-7015 (NOV 1995) (Technical Data) and/or DFARS §227.7202 (Computer Software), as applicable. Contractor/Manufacturer is Environmental Systems Research Institute, Inc., 380 New York Street, Redlands, CA 92373-8100, USA.

Governing Law: This Agreement is governed by the laws of the United States of America and the State of California without reference to conflict of laws principles.

Entire Agreement: The parties agree that this constitutes the sole and entire agreement of the parties as to the matter set forth herein and supersedes any previous agreements, understandings, and arrangements between the parties relating hereto and is effective, valid, and binding upon the parties.

ESRI is a trademark of Environmental Systems Research Institute, Inc., registered in the United States and certain other countries; registration is pending in the European Community. SML, AtlasWare, and Avenue are trademarks of Environmental Systems Research Institute, Inc.

Introduction

Getting started

Using these materials

Unit 1 – The Ocean Basins

Unit 2 – Ocean Currents

Unit 3 – Ocean-Atmosphere Interactions

Unit 4 – Marine Productivity

Introduction

Thinking scientifically

An Earth scientist makes a living by observing and measuring nature. Whether recording and analyzing ocean currents or measuring subtle changes in sea surface temperature over many decades, a successful Earth scientist relies heavily on his or her ability to recognize patterns. Patterns in space and time are the keys to many of the great discoveries about how Earth works. The activities in this module will help you develop your ability to recognize and interpret nature's fundamental patterns by exploring recent scientific data using a computer and geographic information system (GIS) software. GIS provides tools for organizing, manipulating, analyzing, and visualizing information about the world using digital maps and databases.

GIS map

Map of global seafloor age data.

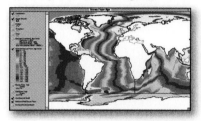

In Unit 1, you will examine the topography and age patterns of the ocean floor to learn about the dynamic processes that shape the ocean basins. In Unit 2, you will explore surface and deep water ocean currents and the factors that influence them. In Unit 3, you will look at energy transfer between the ocean and the atmosphere, and the role it plays in moderating Earth's climate. In the second half of the unit, you will analyze and compare ocean surface conditions and climate patterns during normal, El Niño, and La Niña years. Finally, in Unit 4, you will investigate productivity in the world's oceans and humankind's effects on it.

Tabular data

A tabular form of the seafloor age data. The yellow highlighted rows show data for the Atlantic Ocean.

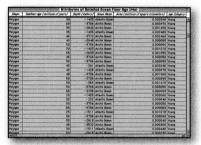

Most of these patterns are presented through maps, which are among scientists' most important tools. Maps allow you to visually explore the spatial relationships between phenomena such as surface winds and ocean currents; natural features such as continents, oceans, and ice caps; and human features such as countries and cities. Behind each map layer or *theme* is a table containing an extensive database of information about each feature in that theme. By carefully analyzing these data, you can reveal patterns in the data that are difficult to discover through visual examination alone. For example, by investigating a series of maps detailing ocean surface properties, you can begin to understand the complex interrelationships that govern the marine food supply.

Planning to learn

Each unit of *Exploring the Ocean Environment* leads you through a well-tested learning process that builds upon your existing knowledge. The first activity of the unit gets you thinking about the major concepts presented in the unit and the key questions that motivate and guide scientific research. It will help you frame your own questions about the topic—questions that you may be able to answer for yourself as you learn more in later investigations.

In the second activity you will explore maps and data looking for patterns. As you examine these patterns, ask questions such as:

- Where do they occur? (or *not* occur?)
- Why does this pattern occur here and not elsewhere?
- What might cause this pattern?

The third activity of each unit provides key background information about major scientific principles and concepts, and should help you begin to answer the questions raised earlier. The readings are brief and explain only the important concepts required to continue or to check your prior answers. In the fourth (and sometimes fifth) activity, you will apply your new knowledge to solve a particular problem. This will help you measure your understanding of the material and prepare you for the final activity where you will be asked to apply the concepts you have learned in the unit to a specific location or situation.

GIS made easier

The purpose of these activities is not simply to learn how to use a GIS, but to use one as a tool to explore and learn about natural processes and features and how they relate to humans and human activities. For this reason, all of the data have been assembled into ready-to-use projects and complex operations have been eliminated or simplified. Although it is helpful to have basic computer skills, you do not need experience with ArcView GIS software to complete these activities. The ArcView user interface has been modified to simplify complex and repetitive processes. Directions for each task are provided in the text, so you will learn to use the tool as you explore with it. The activities barely scratch the surface of the data that have been provided, and we encourage you to explore the data on your own and make your own discoveries.

Getting started

Additional resources

Visit The SAGUARO Project website for updates, references, and links to related websites:

http://www.scieds.com/saguaro

The *Exploring the Ocean Environment* CD-ROM contains all the software and data you need to complete the module activities on your own Macintosh® or Windows® computer.

What you need to know

The authors of this book assume that you know how to use a computer with either the Macintosh or Windows operating system installed. We will make no attempt to teach these basic skills:

- Turning the computer on and, if necessary, logging in as a user
- Navigating the file system to find folders, applications, and files
- Launching applications and opening files
- Using basic interface elements—opening, closing, moving, and resizing windows, using tools, buttons, menus, and dialog boxes

In each activity, you will be instructed to launch the ArcView GIS application, then locate and open a specific file. On your own computer, this file will be located wherever you install it. In a lab setting, your instructor will tell you where to find the file.

Using a lab computer

If you are using this manual as part of a laboratory course, a computer lab with the necessary software and data files may have already been prepared for you in advance. You will not need to install anything, so skip the installation instructions and go directly to the *Using these materials* section on page xv.

Using your own computer

For your convenience, you may install the software and copy the data to your personal computer to complete the exercises in this book. If you are using a Windows-based computer, go to page vi. For Macintosh instructions, go to page ix.

Windows installation

NOTE TO WINDOWS USERS: If you have used another book in the GIS Investigations for the Earth Sciences series (*Exploring the Dynamic Earth, Exploring Tropical Cyclones*, or *Exploring Water Resources*) and installed the version of ArcView that was included with the book, you will not be able to run the version of ArcView provided with *Exploring the Ocean Environment* on the same computer. Feel free to install on a different computer, if one is available. We apologize for this inconvenience, but this is beyond the control of the series authors and publisher.

The instructions below are for installing software on your Windows-based computer. If the software has already been installed, such as in a lab setting, skip ahead to the **Using these materials** section on page xv.

Minimum system requirements

Your computer must meet the following minimum requirements to install and use the software and data for this module. (More and faster is better.)

- 500 MHz or faster Pentium 3-class CPU running Windows 98 or newer. (Windows XP requires patch, included on CD.)
- 128 MB total RAM (64 MB of available application RAM)
- CD-ROM drive
- 700 MB free space on hard drive for applications and data

Required software

- ArcView GIS 3.0–3.3 for Windows (Windows XP requires patch, included.)
- ArcView Dialog Designer extension for Windows (pre-installed with ArcView versions 3.1–3.3, but *not* with the ArcView Virtual Campus Edition that is provided on the *Exploring the Ocean Environment* CD.
- QuickTime™ for Windows 5 or newer
- Acrobat® Reader 4 or newer

You will need to install any software listed above that is not already on your computer. (Note the version numbers. If your version is older, please install the newer version provided. If yours is newer, you may use the version that is already installed.) Installers for all applications and extensions are included on the *Exploring the Ocean Environment* CD-ROM.

Before you install software

- Insert the *Exploring the Ocean Environment* CD into your CD-ROM drive and read the **Readme.rtf** file for last-minute updates.
- Disable virus protection software (if installed) and quit any open applications.
- Be sure you have at least 700 MB of free space on your hard drive.

ArcView 8 / 9 compatibility

ArcView project files in *Exploring the Ocean Environment* are not compatible with ArcView 8 or 9. Data files can be opened by any version of ArcView, 3.0 or newer.

CD-ROM Contents

The **Exploring the Ocean Environment CD** contains the following folders and files.

Readme.txt
License.txt
etoe_unit_1
   etoe_unit_1.apr
   Data (many files)
   Media (many files)
etoe_unit_2
   etoe_unit_2.apr, etc.
etoe_unit_3
   etoe_unit_3.apr, etc.
etoe_unit_4
   etoe_unit_4.apr, etc.
ArcView
   AV32AVC.exe
   dialog.avx
   ArcViewGISPatch4WinXP.exe
Reader
   RP505ENU.exe
QuickTime
   QuickTimeInstaller.exe
Docs
   Guide to ArcView GIS.pdf
   Data Dictionary.pdf
   SAGUARO Tools.pdf

Installing applications

1) ArcView GIS Virtual Campus Edition

- NOTE—This version of ArcView is licensed for 120 days from the first time you use it. You may install it at any time, but do not click the **Try** button on the startup screen until you need to use it, or your license may expire before the semester is over!

- Open the ArcView folder, double-click the **AV32AVC.EXE** icon, and follow the on-screen instructions to complete the installation.

- Restart your computer.

2) Setting Windows compatibility mode

Do this step only if you are using Windows XP home or professional edition on your computer with the ArcView GIS Virtual Campus Edition.

- Locate and open the **ESRI\VCampus\ARCVIEW\BIN32** folder on your hard drive

- Right-click the **ArcView.EXE** icon and choose **Properties** from the contextual menu.

- Click the **Compatibility** tab, then check the **Run this program in compatibility mofe for:** checkbox. Select **Windows 95** from the drop-down menu (see left).

- Click **OK** to close the **Properties** window.

3) Dialog Designer extension

The Dialog Designer extension is required for the *Exploring the Ocean Environment* project files. This extension is pre-installed with ArcView for Windows version 3.1 through 3.3, but not with ArcView 3.0 or the 120-day version of ArcView provided on the *Exploring the Ocean Environment* CD.

- Open the **Dialog Designer** folder found inside the **ArcView** folder of the *Exploring the Ocean Environment* CD.

- Copy the **dialog.avx** file to the **ESRI\VCampus\ARCVIEW\ EXT32** folder on your hard drive.

4) QuickTime

Do this step only if your computer does not have QuickTime for Windows version 5.0 or newer installed.

- Open the QuickTime folder on the *Exploring the Ocean Environment* CD and double-click the **QuickTimeInstaller.exe** icon. Follow the on-screen instructions to complete the installation.

- Restart your computer.

Where is ArcView?

By default, the version of ArcView provided with *Exploring the Ocean Environment* is installed in

C:\ESRI\VCampus\ARCVIEW\BIN32

Why don't I see ".exe" at the ends of any files?

Newer versions of Windows may be set to hide the three-character file extension, so **AV32AVC.EXE** would appear as simply **AV32AVC**.

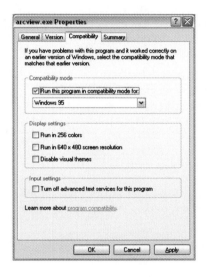

Create a desktop shortcut

You may want to create a shortcut to ArcView on your desktop, to make it easier to access the program. To create a desktop shortcut:

- Right-click the **ArcView** program icon (see note above).
- Choose **Send To > Desktop (create shortcut)** from the contextual menu.

5) Acrobat Reader

If Acrobat Reader (or the newer Adobe Reader) is already on your computer, you may skip this installation.

- Open the Reader folder on the *Exploring the Ocean Environment* CD and double-click the **RP505ENU.EXE** icon. Follow the on-screen instructions to complete the installation.
- Restart your computer.

Installing data

Copy the data files used in this module to your computer's hard drive.

- Copy one or more of the **etoe_unit_x** folders from the *Exploring the Ocean Environment* CD to your hard drive. Each folder corresponds to one unit of the book.
- **Do not change the name of any of the folders or files contained in the etoe_unit_x folder.**
- If you are copying an **etoe_unit_x** folder into another folder, give the enclosing folder a short name (8 characters or fewer) and do not nest the folder too many levels deep. This will help ensure that ArcView can locate the files correctly.
- **Critical issue for Windows installations** - There must be no spaces in the names of the drive or folders along the path to the **etoe_unit_x** folder. If necessary, change spaces in drive and folder names to underscore characters—thus, a Class Data folder should be renamed **Class_Data**.

Setting monitor resolution

The *Exploring the Ocean Environment* module was designed to be used with a monitor resolution of at least 1024 by 768 pixels and 256 colors. To change the monitor resolution under Windows:

- Right-click on the desktop, choose **Settings** from the popup menu.
- Click the **Settings** tab.
- Set the color palette to 256 colors or higher.
- Set the desktop area to 1024 × 768 pixels or higher.

For more help, consult your computer's printed or online documentation.

Installation problems?

Help with common installation and use problems can be found in the *Troubleshooting* section on page xiii.

Installation and setup are complete.
Skip to *Help, Updates, and Resources*, page xii.

Macintosh installation

If you are using a computer in a lab setting where the software and data have already been installed, skip ahead to **Using these materials** on page xv.

Windows users

Skip to **Using these materials** on page xv.

System requirements

Your computer must meet the following minimum requirements to install and use the software and data for this module. (More and faster is better.)

- 500 MHz or faster PowerPC® CPU running Mac OS 8.0 or higher (Under Mac OS X, ArcView requires the Classic operating environment. See **Note to Mac OS X users** below.)
- 128 MB or more total RAM, 64 MB or more available application RAM
- CD-ROM drive
- 700 MB or more free space on hard drive for applications and data

Required software

- ArcView® GIS 3.0a for Macintosh
- ArcView Dialog Designer extension for Macintosh
- QuickTime™ 5 or newer
- Acrobat® Reader 4 or newer

You will need to install any software listed above that is not already on your computer. (Note the version numbers. If your version is older, please install the newer version provided. If yours is newer, you may use the version that is already installed.) Installers for all applications and extensions are included on the *Exploring the Ocean Environment* CD-ROM.

CD-ROM Contents

The *Exploring the Ocean Environment* CD contains the following folders and files.

    Readme.rtf
    License.rtf
    etoe_unit_1
        etoe_unit_1.apr
        Data (many files)
        Media (many files)
    etoe_unit_2
        etoe_unit_2.apr, etc.
    etoe_unit_3
        etoe_unit_3.apr, etc.
    etoe_unit_4
        etoe_unit_4.apr, etc.
    ArcView
        ArcView Installer
        Dialog Designer Installer
    Reader
        Acrobat Reader Installer
    QuickTime
        QuickTime Installer
    Docs
        Guide to ArcView GIS.pdf
        Data Dictionary.pdf
        SAGUARO Tools.pdf

Note to Mac OS X users

Under Mac OS X, ArcView runs in the Classic environment, which provides compatibility with older software. Therefore, you must have Mac OS 9.1 installed on your computer (version 9.2.2 is recommended) in addition to OS X. Newer Macintosh computers do not come with the Classic environment installed by default, and may not even include an OS 9 installation CD. If this is the case, you may obtain an OS 9 installation CD from a friend or users group, your local dealer, or directly from Apple.

The first time you open a Classic application, the Classic environment will take a few moments to start up. Once your application starts up, you'll know you're using a Classic application because the menu bar and windows will have the Mac OS 9 look and feel. See the **Troubleshooting** section on page xiii for additional information about OS X compatibility issues.

Before you install software

- Insert the *Exploring the Ocean Environment* CD into your CD-ROM drive and read the **Readme.rtf** file for last-minute updates.
- Disable virus protection software (if installed) and quit any open applications.
- Be sure you have at least 700 MB of free space on your hard drive.

Intalling applications

1) Install ArcView GIS

ArcView Installer

- Open the **ArcView** folder, open the **Macintosh** folder, and double-click the **ArcView Installer** icon.

- Follow the on-screen instructions to install ArcView on your hard drive. The installer will create a folder named **ESRI** that contains ArcView and its required files at the location you specify.

    OS 9—Install ArcView anywhere on your hard drive.

    OS X—Install ArcView in the **Applications (Mac OS 9)** folder.

- When installation is complete:

    OS 9—Restart your computer and hold down the Command (⌘) and Option keys until the message *"Are you sure you want to rebuild the desktop file on the disk..."* appears. Click **OK**. Your computer will rebuild its desktop file and complete the startup process.

    OS X—Quit the installer. The Classic environment will quit and restart automatically.

2) Increase ArcView's memory allocation

The default memory allocation for ArcView must be increased to assure trouble-free operation! Follow these steps to increase the application's memory allocation.

- Navigate to where you installed the ArcView GIS application. Look for a folder named **ESRI**. Open it, then open the folder inside it named **AV_GIS30a**.

- Single-click the **ArcView** application icon to select it.

- Choose **File > Get Info**.

    OS 9—choose **Memory** from the **Show** pop-up menu.

    OS X—click the triangle to expand the **Memory** section.

- Enter at least **32000** for the **Minimum Size** and **64000** for the **Preferred Size**. (Do not include commas!) If your computer has more memory (RAM, not hard drive space) available, you can allocate some of the additional memory to ArcView, keeping in mind your other system requirements.

- Close the **ArcView Info** window.

3) Register ArcView GIS

To register the ArcView GIS application:

- Double-click the ArcView icon to launch the application.

- When prompted, enter your name and company (you may leave this blank), then enter the registration number: **708301184404**.

- Choose **File > Exit** to quit ArcView.

Make an alias!

For convenience, you may wish to make an alias of the ArcView application on your desktop.

- Click the application icon to select it.
- Choose **File > Make Alias** to create an alias icon.
- Drag the alias icon onto the desktop.

Increasing memory

Under Mac OS 9

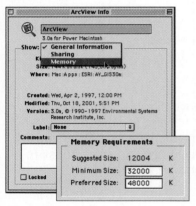

Under Mac OS X

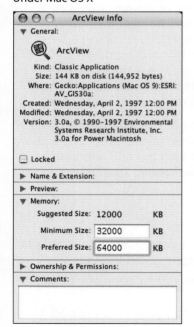

4) Install Dialog Designer extension

Dialog Designer is an ArcView add-on that provides the ability to modify the ArcView user interface to make the program easier to use. It is required for the *Exploring the Ocean Environment* project files.

- Open the **ArcView** folder on the CD, then open the **Dialog Designer** folder.
- Double-click the **Dialog Designer Installer** icon   to launch the installer.
- Navigate to the **AV_GIS30a** folder in the **ESRI** folder as directed, and install the dialog designer extension. When it has finished, quit the installer.

5) Install QuickTime

The investigations in this module use QuickTime to display movies and animations. Most Macintosh computers already have QuickTime installed. If yours does not, you should install it.

- Open the **QuickTime** folder on the CD and double-click the **Quicktime Installer** icon. Follow the on-screen instructions to complete the installation.
- Restart your computer.

6) Install Acrobat Reader (optional)

Acrobat Reader is used to view and print the files in the **Docs** folder. If Acrobat Reader is already installed on your computer or you do not wish to view or print these documents, you may skip this installation.

- Open the **Reader** folder on the CD and double-click the **Acrobat Reader Installer** icon. Follow the on-screen instructions to complete the installation.
- Restart your computer.

Installing data

Copy the data files used in this module to your computer's hard drive.

- Insert the *Exploring the Ocean Environment* CD in your CD-ROM drive. Double-click the CD icon on the desktop to view the contents of the disc.
- Copy one or more of the **etoe_unit_x** folders from the CD to your hard drive. Each folder corresponds to one unit of the book.
- **Do not change the name of any of the folders or files contained in the etoe_unit_x folder.**
- If you are copying an **etoe_unit_x** folder into another folder, give the enclosing folder a short name (8 characters or fewer) and do not nest the folder too many levels deep. This will help ensure that ArcView can locate the files correctly.
- If the data files get renamed or damaged, delete the entire **etoe_ unit_x** folder and copy a clean version of the folder from the CD to your hard drive.

Setting monitor resolution

Exploring the Ocean Environment was designed to be used with a monitor resolution of at least 1024 by 768 pixels and 256 colors. To change the monitor resolution under Mac OS 8/9:

- Choose **Control Panels ▶ Monitors** under the Apple Menu.
- Click the **Monitor** button.
- Set the **Resolution** to **1024 × 768** pixels or higher, and the **Colors** to **Thousands** or **Millions**.

Under Mac OS X:

- Choose **System Preferences...** under the Apple Menu.
- Click the **Displays** button.
- Set the **Resolution** to **1024 × 768** pixels or higher, and the **Colors** to **Thousands** or **Millions**.

For more help, consult your computer's documentation.

Help, updates, and resources

For corrections, updates, and additional resources related to *Exploring the Ocean Environment*, visit the SAGUARO Project website at:

http://www.scieds.com/saguaro/etoe/

Two Help menus on a Mac?

On the Macintosh, ArcView adds a second Help menu of its own to the left of the standard Macintosh Help menu. Use this menu to access ArcView's built-in help system.

Online help

This module provides all of the directions you need to complete the activities using ArcView GIS. If you wish to explore ArcView on your own, or learn more about ArcView's capabilities, you may wish to consult ArcView's Help menu. Choose **Help > How to Get Help...** to learn more about using ArcView's built-in help system.

Printing the pdf files in the Docs folder

The **Docs** folder contains the following reference materials, which you can view or print using the included Acrobat Reader software.

- **Guide to ArcView GIS.pdf**—a quick reference manual to commonly used ArcView tools and techniques.
- **SAGUARO Tools.pdf**—explains the interface and usability enhancements to ArcView developed by the SAGUARO Project and built into the *Exploring the Ocean Environment* project files.
- **Data Dictionary.pdf**—provides information about the data used in this module, including a description of where the data came from and how they were processed. You do not need this information to complete the activities, but it may be useful if you are using the data for independent research.

Troubleshooting

Q&A about the 120-day ArcView license on the CD

"My university has an ArcView site license. Do I need to use the 120-day version of ArcView on the Exploring the Ocean Environment CD?"

Only if you want to use your own computer to complete the GIS activities. The 120-day version of ArcView is primarily intended for use on students' personal computers.

"When should I install the 120-day version of ArcView?"

Do not install your ArcView GIS software until you need to use it. The ArcView license allows you to install the ArcView application on one computer for 120 days. At the end of that time, the software will no longer function, and it cannot be reinstalled on that computer.

"Can I reinstall the application if it gets damaged or someone accidentally uninstalls it?"

Yes, but it will function for only 120 days from the original installation date.

"What happens at the end of the 120-day license period?"

Your copy of ArcView GIS will stop working. According to the license agreement, you must uninstall the ArcView GIS application, because it will no longer function and will not work if reinstalled. For pricing and information about student and instructor licenses for ArcView, call ESRI telesales at (800) 447-9778.

"What happens if I retake this course or take a different course using another module from the Exploring series?"

You must install your "new" 120-day licensed version of ArcView on a different computer than your original installation.

"When I _____ in ArcView, nothing (or the wrong thing) happens."

Most ArcView errors are caused by not having the correct theme activated when performing an operation. By activating a theme, you are telling ArcView which data to operate on. If the wrong data are identified, the operation will most likely fail or produce unexpected results.

"ArcView crashes while launching." (Macintosh and Windows)

- ArcView may not have enough memory to work properly (Macintosh only.) See page x for directions on increasing ArcView's memory allocation.
- ArcView may be damaged. You may uninstall and reinstall ArcView at any time within the 120-day license period using the *Exploring the Ocean Environment* CD.

"When I open a project file, ArcView keeps telling me it cannot find files that I know are installed." (Macintosh and Windows)

The project file contains a path to the location of each data file in the project. If you move files or rename any of the files or folders that were installed, ArcView cannot find the files. If you cannot restore the correct file and folder names and locations, delete the **etoe_unit_x** folder and re-install it on the hard drive.

"When I open a project file from within ArcView, it tells me it can't find the project file." (Windows only)

There is probably a space in the name of the drive or folder into which you copied the **etoe_unit_x** folder. ArcView for Windows cannot locate a file correctly if it encounters a space in the file's pathname. If necessary, change spaces in drive and folder names to underscore characters—thus, a folder named **Class Data** should be renamed **Class_Data**.

"ArcView launches and opens the project file, but it crashes later, after several minutes of use ." (Macintosh)

The memory allocation for ArcView is probably too low. See page x for directions on increasing ArcView's memory allocation.

"When I try to view a QuickTime movie or animation, I get the message 'Required compressor not found.'" (Macintosh and Windows)

This message means that you are using an older version of QuickTime that does not support the data compression format used for the movie. Install QuickTime 5 or newer to correct this problem.

"QuickTime movies don't play smoothly." (Macintosh and Windows)

The quality of QuickTime movie playback depends on many factors. Quit any unnecessary applications that are running, and be sure you are using QuickTime 5 or newer.

"When I use the Media Viewer to view movies or animations, I get a message inviting me to upgrade to QuickTime Pro. When I click the 'Later' button, nothing happens." (Windows only)

This usually occurs only in a lab or other shared environment, when the QuickTime preferences file is not "writable" to all users. The preferences file is called **QuickTime.qtp**; its exact location depends on the version of Windows you are using. Consult your system administrator about making the necessary changes.

"When I use the Media Viewer to load an animation or web page, ArcView freezes." (Mac OS X only)

ArcView runs in the Macintosh Classic environment, whereas QuickTime Player and your web browser both run natively under OS X. When you launch a web page or animation, ArcView waits for a confirmation that the correct software has answered its request to open the file. For some reason, ArcView does not always receive or acknowledge this confirmation, even though the file opens properly. ArcView "gives up" after waiting about a minute, and resumes its normal behavior. If you are patient, the "freeze" will clear in about 60 seconds, allowing you to continue the investigation. We apologize for this bug, but it appears to be beyond our control!

"When I use the Identify tool 🛈 or perform certain operations, ArcView gives me an error message or crashes." (Windows XP only)

If you are using the *ArcView Virtual Campus Edition* that was included on your *Exploring the Ocean Environment* CD, you must run the application in Windows 95 compatibility mode. See step 2 on page vii for detailed instructions.

If you are using any other version of ArcView 3.X, you must update it using the Windows XP patch application, **ArcViewGISPatch4WinXP.EXE**, located in the ArcView folder on your *Exploring the Ocean Environment* CD-ROM.

"When I open a project file, ArcView crashes or I get an error message saying that the Dialog Designer extension cannot be found." (Macintosh and Windows)

Dialog Designer is a free ArcView extension that is required by all project files in *Exploring the Ocean Environment*. Quit ArcView and follow the instructions on page vii (Windows) or xi (Macintosh) to install the appropriate version of the Dialog Designer extension for your computer's operating system.

Using these materials

Visual cues and basic skills

Visual cues are used to make the activity directions easier to follow.

- A line preceded by the ▶ symbol is an instruction—something to do on the computer.
- When referring to a tool or button, the name of the tool or button is capitalized and is followed by a picture of that item as it appears on screen—for example, the Identify tool 🅸.
- The > symbol between two boldface words in text indicates a menu choice. Thus, **Theme > Properties** means "pull down the Theme menu and choose Properties."

Theme > Properties *means...*

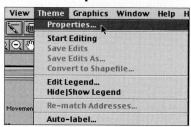

Sidebar contains important information!

The sidebar on each page contains useful information such as definitions, explanations, illustrations, examples, reminders, warnings, tips, and hints. If you are not sure what to do, the first place to look is the sidebar!

Activating themes

Most common problems encountered while using ArcView GIS software are caused by failing to activate the correct theme (map layer) before performing an operation. In most cases the instructions specifically note when to activate a theme, but users often skip over this detail, or confuse *activating a theme* with *turning the theme on*.

To activate a theme, click on its name in the Table of Contents. Active themes are indicated by a raised border.

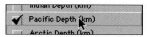

To activate a theme, single-click on the theme's name in the Table of Contents, the list of themes to the left of the map view. Active themes are indicated by a raised border in the Table of Contents. When things do not seem to be going as they should, this is the first thing to check.

Zooming

The second most common source of problems is not looking closely enough at the maps to see the necessary feature(s). ArcView has tools for zooming in and out of the map view that work just like tools you have used in other applications. If you zoom in so far that you don't know where you are, you can back up to previous zooms by clicking the Previous extent button ⬅. You can also zoom out quickly to see the entire map by clicking the Full Extent button 🌐 or by choosing **View > Full Extent**.

To zoom in on an area, click and drag with the Zoom In tool 🔍 to outline the area. To zoom out, click anywhere on the map window with the Zoom Out tool 🔍.

Want to know more?

If you would like more information on how to use ArcView GIS, refer to the **Guide to ArcView GIS.pdf** and **SAGUARO Tools.pdf** files in the **Docs** folder on the *Exploring the Ocean Environment* CD.

Large numbers

Some of the numbers you will work with in *Exploring the Ocean Environment* are quite large. When talking about the amount of water in the ocean or the area of an ocean basin, you routinely use values in the billions or even trillions! Where possible, ArcView has been modified to make these very large and very small numbers easier to read. For example, in the Statistics window shown at left, the total area is given as 349.96 trillion, rather than 349,958,342,077,361 square meters.

Occasionally, you will need to convert billions to trillions or millions, or vice versa. For example, to convert the **Mean** value in the example at left from billions to trillions, move the decimal point three places to the *left*. To go from billions to millions, move the decimal three places to the *right*.

14490 million = 14.49 billion = 0.01449 trillion

Statistics for Area (m^2) [Phosph

Statistics Report: Phosphates (μM)
============================
AREA (M^2): ----------------
  Total: 349.96 trillion
  Mean: 14.49 billion
  Minimum: 107.90 million
  Maximum: 1.75 trillion
  Number of Features: 24,149

Rounding examples

For example, if your number is

319,740,562.85

To round to the nearest ten million:

- Find the ten millions digit (1).
- Look at the number to its right (9). Because it is between 5 and 9, add one to the ten millions digit.
- Change the whole numbers to the right of the ten millions digit to zeros. The result is **320,000,000**.

To round to the nearest 0.1 million:

- Find the 0.1 millions digit (7). (This is also called the hundred thousands digit.)
- Look at the number to its right (4). Because it is between 0 and 4, do not add one to the 0.1 millions digit.
- Insert the decimal point in the proper location. The result is **319.7 million**.

Rounding this number to the nearest...

...million (1,000,000) = **320,000,000** (adding 1 to 319 gives 320)
...hundred thousand (100,000) = **319,700,000**
...ten thousand (10,000) = **319,740,000**
...thousand (1,000) = **319,741,000**
...hundred (100) = **319,740,600**
...ten (10) = **319,740,560**
...one (1) = **319,740,563**
...tenth (0.1) = **319,740,562.9**

Rounding

Most of these numbers are approximations, so it does not make sense to be overly precise when you are calculating or recording them. Look at the number written below, and the values of each of the digits. Face it—when you are talking about nearly 149 billion of something, who cares about hundred-thousandths, or even tens of millions?

148,735,992,068.95249

Throughout Exploring the Ocean Environment, you will be asked, for example, to "Round your answer to the nearest 0.1 million." Rounding numbers is very simple, if you follow these three steps:

- Look only at the numeral to the right of the place value you are rounding to. For example, when rounding to the nearest thousand, look only at the numeral in the hundreds place.
- If the numeral to the right is 0-4, do not change the number in the rounding place value. If the number to the right is 5-9, add one to the number in the rounding place value.
- Change the whole numerals to the right of the place you are rounding into zeros, and omit all decimal places and the decimal point.

Rounding decimals

Rounding decimals works exactly the same way as rounding whole numbers. The only difference is that instead of rounding to tens, hundreds, thousands, and so on, you round to tenths, hundredths, thousandths, and so on. Do not add zeros to the right of the decimal point. In other words, rounding 2.587 to the nearest tenth is 2.6, *not* 2.600.

Estimating percent area

You will occasionally be asked to estimate the percent area covered by land, ocean, or some other feature. This is a difficult skill for some people to master, but can be learned with practice.

Comparing to standards

One method of estimating coverage is to compare to visual standards. As you can see from the examples below, when estimating coverage you need to consider how the features are arranged.

Cloud cover exercise

Here is a simple activity that demonstrates the confusing nature of cover estimates.

- Take two full sheets of blue paper and one of white paper. The blue paper represents sky, and the white paper represents clouds.
- Cut the white sheet in half. Tear or cut the first half of the white sheet into large pieces and glue them onto one of the blue sheets without overlapping.
- Repeat with the other half of the white sheet and the other blue sheet. This time, cut or tear the white sheet into small chunks before gluing them on.

In both cases, the cloud cover is 50%. Half of the blue sky is covered by white clouds, but the sheet covered by large clouds appears more open than the sheet covered by small clouds.

Random		**Grouped**
	0%	
	10%	
	20%	
	30%	
	40%	
	50%	
	60%	
	70%	
	80%	
	90%	
	100%	

Gridding

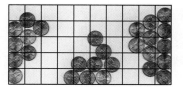

Another approach to estimating coverage is to divide the area up into a grid, either mentally or physically, and determine the number of grid squares that are at least half-covered. To find the percent coverage, calculate the ratio of covered squares to total squares and multiply by 100.

In the example at left, approximately 20 of the 50 squares are at least half covered.

$$20/50 \times 100 = 40\% \text{ coverage}$$

Using ArcView GIS for Windows

Launching ArcView GIS and opening a project file

• Double-click the ArcView shortcut icon on the desktop (if you created one) or click the **Start** button on the Windows Taskbar. On the **Start** menu, choose **All Programs > ESRI > ArcView GIS [version or edition] > ArcView GIS [version or edition]**.

Which version of ArcView should I use?

You can use *ArcView GIS 3.0–3.3* or the *ArcView GIS Virtual Campus Edition* (based on version 3.2) with these materials. If more than one version exist on your computer, use the newest version available. All available versions should be listed on the **Start > All Programs > ESRI** pop-up list.

• **If you are *not* using the 120-day Virtual Campus Edition of ArcView GIS, skip to the next step.** On the Virtual Campus Edition startup screen, click the **Try** button. This will begin your 120-day license period. Each time you launch the ArcView GIS application, this screen tells you the number of days remaining in the license period.

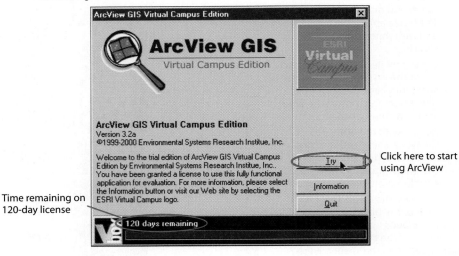

Click here to start using ArcView

Time remaining on 120-day license

Welcome to ArcView GIS

Open project files for *Exploring the Ocean Environment* activities from the **Welcome to ArcView GIS** dialog box.

• When the **Welcome to ArcView GIS** dialog box appears (see left), click the **Open an existing project** button then click **OK**.

• Use the **Open Project** dialog box to locate the file you want to open.

1) Select the drive where you installed the module files.

2) Select the directory (folder) containing the project files.

3) Select the project file. ArcView project files end in **.apr**.

4) Click **OK**.

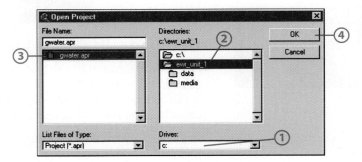

- When the project file opens, you may see the following message. Answer either **Yes** or **No**—you do not need to load the new ArcView tools, but they will not harm anything if you choose to load them.

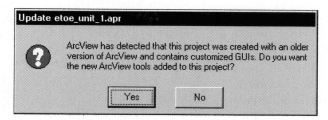

Closing a project file

When you have completed an activity or must stop for some reason,

- Choose File > Exit.

- When asked if you want to save your changes, click **No**. (Don't worry if you click **Yes**. The files have been locked to prevent accidentally modifying or erasing them. If you should somehow mess up the original file, you can replace it with a fresh copy from the *Exploring the Ocean Environment* CD.)

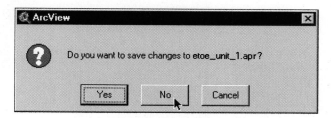

Using ArcView GIS for Macintosh

Launching ArcView and opening project files

Where is the ArcView icon?

If you created an alias of the ArcView application icon on your desktop, double-click it to launch ArcView. Otherwise, open the **ESRI** folder that was installed on your hard drive when you installed ArcView, and double-click the ArcView icon located in the **AV_GIS30a** folder.

- Double-click the **ArcView** application icon to launch ArcView GIS. (See note at left.)

- Choose **File > Open....**

- Navigate to the appropriate **etoe_unit_x** folder installed on your hard drive, select the ArcView project file (the file name ends with **.apr**), and click **Open**.

Closing project files

When you have completed an activity or must stop for some reason,

- Choose **File > Exit**.

- When asked if you want to save your changes, click **No**. (Don't worry if you click **Yes**. The files have been locked to prevent accidentally modifying or erasing them.)

The ArcView GIS user interface

Basic elements

The basic elements of the ArcView interface are essentially identical for the Windows and Macintosh versions, except for the location of the Status Bar.

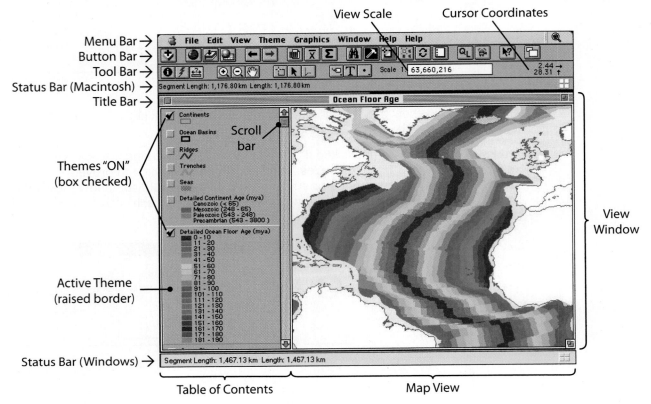

Menu Bar →
Button Bar →
Tool Bar →
Status Bar (Macintosh) →
Title Bar →

View Scale   Cursor Coordinates

Themes "ON" (box checked)

Scroll bar

Active Theme (raised border)

Status Bar (Windows) →

View Window

Table of Contents   Map View

Buttons and tools

In *Exploring the Ocean Environment,* ArcView has been modified to group related buttons and tools and to simplify frequently-used operations.

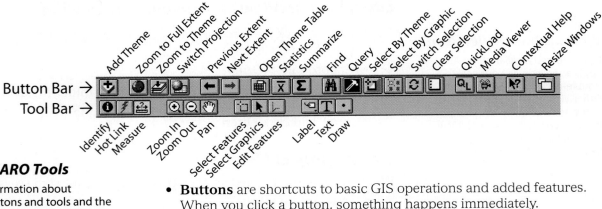

Button Bar →
Tool Bar →

Add Theme, Zoom to Full Extent, Zoom to Theme, Switch Projection, Previous Extent, Next Extent, Open Theme Table, Statistics, Summarize, Find, Query, Select By Theme, Select By Graphic, Switch Selection, Clear Selection, QuickLoad, Media Viewer, Contextual Help, Resize Windows

Identify, Hot Link, Measure, Zoom In, Zoom Out, Pan, Select Features, Select Graphics, Edit Features, Label, Text, Draw

The SAGUARO Tools

For more information about ArcView's buttons and tools and the usability enhancements developed by the SAGUARO Project, see the **SAGUARO Tools.pdf** file in the **Docs** folder of the *Exploring the Ocean Environment* CD.

- **Buttons** are shortcuts to basic GIS operations and added features. When you click a button, something happens immediately.

- **Tools** allow you to interact with map features and edit graphics. To use a tool, click on its toolbar icon, then click or drag on the Map View. The cursor changes to indicate the tool you are using.

Unit 1
The Ocean Basins

In this unit, you will...

- *Track changes in the world's continents and ocean basins from 750 million years in the past to 250 million years in the future*
- *Contrast the extent and age distribution of oceanic and continental crust and hypothesize about their origins*
- *Investigate the origin of deep ocean basins*
- *Create bathymetric profiles of an ocean basin at various resolutions*
- *Examine surface features that reveal the dynamic nature of the ocean basins*

Surface features of ocean basins show that the ocean floor is constantly changing.

Activity 1.1

The once and future ocean

Earth is unique in our solar system for its vast oceans of water. With over 70% of its surface covered by water, Earth is often called *the water planet*. Without this water, Earth would not have the diversity of life forms, the continents we live on (it takes water to create most continental rocks), or the atmosphere that protects us. Have you ever wondered where this water came from or how and when the ocean basins formed? In this unit, you will explore these questions. To start, think about these questions and write your best answers based on your current understanding of Earth.

Oceans on other planets?

In early 2004, twin NASA rovers *Spirit* and *Opportunity* landed on Mars and found evidence that large quantities of water existed on its surface in the distant past. How would an ocean on Mars have been different from one on Earth, and what caused it to disappear?

1. Where do you think Earth's vast supply of water came from? Explain.

If Earth were flat, with no mountains or basins, water would cover the entire planet.

2. What might cause Earth to have deep basins surrounded by high continents to hold the ocean water?

The eternal ocean

And thou vast ocean, on whose awful face
Time's iron feet can print no ruin-trace.

Robert Montgomery, *The Omnipresence of the Deity*

The sun's lifetime

Astronomers estimate that the Sun's life span will be about 10 billion years, and that it is currently around 4.6 billion years old.

Poets and writers are fond of saying that the oceans are eternal, that a stroll along a sandy beach provides a glimpse into Earth's past and a look into its future. True, the oceans have been around for a long time—at least 3.8 billion years. They will probably be around until the Sun expands and boils them away billions of years in the future. However, have they changed at all over time? For now, consider the poet's statement that the oceans are eternal and unchanging. How true is this? You will use ArcView GIS to explore this question.

Moving continents

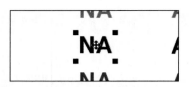

Figure 1. Selecting graphic labels. Four small black squares or *handles* appear around the label when it is selected.

To turn a theme on or off, click its checkbox in the Table of Contents.

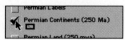

What are Ma and m.y.?

The Greek prefix Mega (symbol = M) represents *one million* in the metric system. The Latin word for year is *annum* (symbol = a). As used in Earth science, Ma stands for *Mega annum*, or *millions of years before present*.

The abbreviation m.y. stands for *millions of years*. A + sign before the number represents millions of years *in the future*. (For example, +250 m.y. means *250 million years in the future*.)

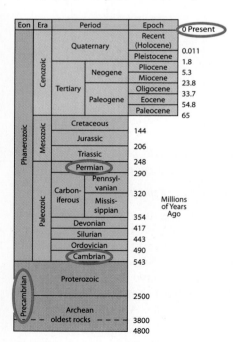

Figure 2. The geologic time scale. Note that the time divisions are **not** to scale!

▶ Launch ArcView GIS, then locate and open the **etoe_unit_1.apr** project file.

▶ Open the **Once and Future Oceans** view.

This view shows the present, familiar locations of today's continents and oceans. However, the continents and oceans have not always been in these locations, nor will they be in the future. In this investigation, you will trace the movements of three continents from 750 million years ago to 250 million years in the future.

▶ Using the Select Graphics tool 🔧, click on the black North America label (*NA*) in the row of Present labels (Figure 1).

▶ Drag the selected label and place it in the middle of the continent it represents (North America).

▶ Turn off the **Present Land**, **Present Continents**, and **Present Labels** themes, and turn on the four **Permian** themes (Oceans, Land, Continents, and Labels).

During the Permian period, around 250 million years ago (Figure 2), the world looked very different than it does today. The **Permian Continents (250 Ma)** theme shows the approximate boundaries of the continents during the Permian period.

▶ Drag the North America (NA) label for the Permian period (250 Ma) to the approximate center of that continent during that period.

▶ Turn off all of the **Permian** themes.

▶ Repeat this process to plot the location of North America for the **Cambrian (500 Ma)**, **Precambrian (750 Ma)**, and **Future (+250 m.y. in the future)** themes.

▶ Turn on the **Present Land** theme and turn all other themes off.

3. On Map 1, mark and label the location of North America during each time period. Draw a line with arrows showing the changing location of North America, from Precambrian to Cambrian to Permian to Present to Future.

▶ Drag the labels out of the map area before you begin plotting the path of the next continent.

▶ Repeat this process to plot the motions of Africa (Af) and Australia (Au) on Map 1. Plot each path using a different color line or pattern. (Hint: When you plot the path of Australia, remember that the left edge of the map connects with the right edge of the map because Earth is round. So, your path may have to break when it reaches the edge of the map.)

Map 1—Moving continents (750 Ma to +250 m.y.)

4. In Table 1, describe the motion of each continent, in terms of its direction and speed, using the words *constant* or *changing*. Because the time intervals are equal—250 million years—the distance between labels is directly related to the speed of motion.

Table 1—Motion of continents over time

Continent	Direction	Speed
North America (NA)		
Africa (Af)		
Australia (Au)		

▶ You are finished with the labels. To delete them, choose **Edit > Select All Graphics** followed by **Edit > Delete Graphics**. (If parts of the labels still appear, they will disappear in the next step.)

▶ Turn on the **Present Land** and **Present Atlantic Ocean** themes and turn all other themes off.

The changing Atlantic

Next, you will focus on changes in the Atlantic Ocean over time. The **Present Atlantic Ocean** theme outlines the Atlantic Ocean as it exists today, situated between North and South America to the west and Europe and Africa to the east.

▶ Turn off the **Present Atlantic Ocean** and **Present Land** themes and turn on the **Permian Land (250 Ma)**, **Permian Continents (250 Ma)**, and **Permian Oceans (250 Ma)** themes.

The view now shows the world's land and ocean areas during the Permian period, 250 million years ago.

5. In Table 2, indicate whether an ocean (an earlier Atlantic Ocean) did or did not exist during the Permian period.

Moving continent animations

To view animations of the continents moving over time, click the Media Viewer button 🎛 and choose either of the following movies from the media list.

- **Atlantic Basin History**
  Shows the changes in the Atlantic Ocean basin.
- **Moving Continents**
  Shows what Earth may have looked like as the continents moved.

Table 2—The Atlantic Ocean basin (250 Ma to +250 m.y.)

Time	Permian (250 Ma)	Present	Future (+250 m.y.)
Does the Atlantic Ocean exist?		exists	

▶ Turn off the **Permian** themes and turn on the **Future Land (+250 m.y.)**, **Future Continents (+250 m.y.)**, and **Future Labels** themes.

This map shows one possible arrangement of the continents 250 million years in the future. While the past locations and extents of continents are fairly well known, future locations are highly speculative. The locations shown here represent just one of several possible scenarios for the future Atlantic Ocean basin.

6. In Table 2, indicate whether this model suggests that the Atlantic Ocean will exist or will not exist 250 million years in the future.

7. Based on Table 2, describe what happens to the Atlantic Ocean basin from 250 million years ago to 250 million years in the future.

This investigation began with a quote from the poet Robert Montgomery that suggested that the oceans are eternal and unchanging.

8. Why do you think the oceans *seem* eternal?

9. Based on what you have learned, give a brief statement that argues *against* the notion that oceans are eternal.

Throughout this unit, you will explore the evidence for and the driving forces behind the motion of the continents and the opening and closing of ocean basins. In later units, you will be challenged to think about the effects these changes have on ocean and atmospheric circulation, climate, marine productivity, and biological evolution and diversity.

▶ Choose **File > Exit** to quit ArcView GIS. Do not save your changes.

Activity 1.2

Changing oceans

Oceans have existed on Earth for at least 3.8 billion years. It is easy to think that they are permanent features, ancient and unchanging. However, the only thing that is truly constant about ocean basins—and continents—is that they are always moving and changing. New oceans are born and existing oceans change size and shape or disappear entirely. How does this happen, and over what time scale? In this activity, you will explore evidence to answer these questions.

Using seafloor age data, you will identify major transformations that have occurred in Earth's ocean basins and when they took place. Differences in the ages of the ocean basins and continents provide clues to understanding when and how they formed.

Ocean floor age

By measuring the ages of ocean floor rocks both directly and indirectly, scientists have produced age maps of most of the ocean basins.

▶ Launch ArcView GIS and locate and open the **etoe_unit_1.apr** project file.

▶ Open the **Ocean Floor Age** view.

What are Ma and m.y.?

The Greek prefix Mega (symbol = M) represents *one million* in the metric system. The Latin word for year is *annum* (symbol = a). As used in Earth science, Ma stands for *Mega annum*, or *millions of years before present*.

The abbreviation m.y. stands for *millions of years*. A + sign before the number represents millions of years *in the future*. (For example, +250 m.y. means *250 million years in the future*.)

The **Detailed Ocean Floor Age (Ma)** theme shows the ages of the rocks that form the present ocean floor. Each colored band represents a time span of 10 million years.

1. Use the legend for the **Detailed Ocean Floor Age (Ma)** theme to complete Table 1.

Table 1—Ocean floor age

Rocks	Color	Age (Ma)
Youngest		
Oldest		

2. Describe the pattern of rock ages on the ocean floor. Are the ages uniform, do they increase or decrease in a pattern, or are the ages just randomly distributed?

To turn a theme on or off, click its checkbox in the Table of Contents.

To activate a theme, click on its name in the Table of Contents.

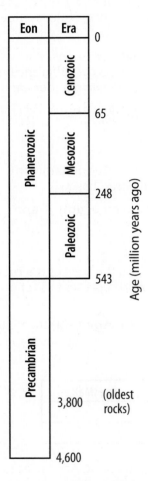

Figure 1. Geologic time scale. Time divisions are *not* drawn to scale.

▶ Turn on the **Detailed Continent Age (Ma)** theme.

This theme displays the ages of surface rocks on the continents, in increments that correspond to the major eras of geologic time (Figure 1).

3. Complete Table 2 using the legend for the **Detailed Continent Age (Ma)** theme.

Table 2—Continental rock age

Rocks	Color	Era / Eon	Age (Ma)
Youngest			
Oldest			

4. Does the age distribution of rocks on the continents follow a pattern similar to the one you saw in the oceanic rocks? Explain.

5. How many times older are the oldest continental rocks than the oldest rocks that form the ocean floor? (Hint: Divide the age of the oldest continental rocks in millions of years by the age of the oldest ocean floor rocks in millions of years.)

Comparing oceanic and continental rocks

Next, you will compare the percentages of the ocean floor and the continents that are covered by young and old rock. You will use simplified versions of the ocean floor and continental age themes that categorize the rock as either young, created in the Cenozoic era (0-65 Ma) or old, created prior to the Cenozoic era (65-3800 Ma).

▶ Turn off the **Detailed Ocean Floor Age (Ma)** and **Detailed Continent Age (Ma)** themes and turn on the **Ocean Floor Age** and **Continent Age** themes.

Now you will determine the percentage of young rocks in the ocean floor and the continents.

▶ Click the Query button ![icon], then follow steps 1-6 below to enter the query (**[Age Category] = "young"**). In step 1, select the **Ocean Floor Age** theme.

1) Select Theme   2) Double-click Field   3) Single-click operator   4) Type Value in quotes   5) Set Highlight   6) Click New

Query Builder

Themes
- Detailed Continent A
- Detailed Ocean Floor
- Ocean Floor Age
- Continent Age
- Continental Shelf
- Unclassified Ocean F

Fields
- [Age (Ma)]
- [Depth (m)]
- [Ocean Basin]
- [Area (million km^2)
- [Age Category]

Operators: = <> and > >= or < <= not ()

Values

☐ Update Values

Query Statement:
([Age Category] = "young")

New
Add
Subtract
Subset

Current Filter Definition:
– No Filter Definition Specified –

Clear Highlighted

QuickLoad Query

Found Features:
○ Display only   ● Highlight

Cancel

Number of features highlighted = 6444

QuickLoad Query

If you have difficulty entering the query statement correctly:

- Click the **QuickLoad Query** button in the Query Builder dialog box and load the **Young Ocean Floor** query.
- Select the **Highlight** option.
- Click **New**.

▶ Close the Query Builder window.

The younger rocks of the ocean floor should now be highlighted yellow.

▶ Click the Statistics button ![icon] to calculate statistics for **the selected features of** the **Ocean Floor Age** theme, using the **Area (million km^2)** field. Select the **Basic** output option and click **OK**. (Be patient—ArcView may take a while to calculate statistics.)

▶ The total area of the ocean floor that is covered by young rock is reported as the **Total**.

6. Round the total area of young ocean floor rock to the nearest whole number and record it in Table 3.

Calculate Statistics

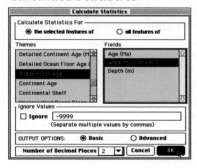

Unclassified?...

The first row in Table 3 is labeled *unclassified,* and this information is provided for you. The unclassified areas represent the regions of the ocean floor and continents where the age of the rock has not yet been determined. This information was included to allow you to determine the *total* area of the oceans and the continents.

Table 3—Area and percent area of oceanic and continental rocks

Age (million yrs)	Ocean Floor		Continents	
	Area (millions km²)	% Area	Area (millions km²)	% Area
unclassified	32		2	
young (0-65)				
old (65 & older)				
Total		100		100

▶ Close the statistics window.

▶ Activate the **Ocean Floor Age** theme.

▶ Click the Switch Selection button 🔁 to switch your selection from the young to the old ocean floor rock. The old ocean floor rock should now be highlighted yellow.

▶ Click the Statistics button 🗴 and calculate statistics for **the selected features of** the **Ocean Floor Age** theme, using the **Area (million km^2)** field. Select the **Basic** output option and click **OK**.

7. Round the total area of old rock in the ocean floor to the nearest whole number and record it in Table 3.

▶ Close the statistics window.

▶ Click the Clear Selection button 🔲.

▶ Activate the **Continent Age** theme.

▶ Repeat these steps to obtain the area of the *continents* covered by young and old rock.

- Query the **Continent Age** theme for (**[Age Category] = "young"**), and click **New**.

- Click the Statistics button 🗴 and calculate statistics for **the selected features of** the **Continent Age** theme, using the **Area (million km^2)** field. Select the **Basic** output option and click **OK**.

- Round the total area of young continental rock to the nearest whole number and record it in Table 3.

- Click the Switch Selection button 🔁. This will select and highlight the old continental rock.

- Click the Statistics button 🗴 and calculate statistics for **the selected features of** the **Continent Age** theme, using the **Area (million km^2)** field. Select the **Basic** output option and click **OK**.

- Round the total area of old continental rock to the nearest whole number and record it in Table 3.

You can use this information to determine the percentage of ocean floor and continents represented by each age category.

8. Add the areas of unclassified, young, and old ocean floor rock and record the total area in Table 3. Repeat this process to record the total area of continental rock.

9. Calculate and record the percentages of unclassified, young, and old ocean floor rock, and record your results in Table 3. Repeat this process for the continental rock. Table 3 should now be complete.

10. According to Table 3, which contains a larger percentage of young rock (less than 65 Ma)—the ocean floor or the continents?

QuickLoad Queries

If you have difficulty entering the query statements correctly in the Query Builder dialog box:

- Click the **QuickLoad Query** button in the Query Builder dialog box.
- Load the **Young Continent** query.
- Select the **Highlight** option.
- Click **New**.

Calculating percentages

Example: To calculate the percent of unclassified ocean floor rock, divide the unclassified area by the total area and multiply by 100.

Age (million yrs)	Ocean Area (millions km²)
unclassified	(32)
young (0-66)	
old (66 & older)	
Total	(306)

% Unclassified Ocean Floor

$$= \frac{32 \text{ million km}^2}{306 \text{ million km}^2} \times 100 = \textbf{10.5\%}$$

Graph 1 shows the amount, in square kilometers, of rocks of different ages in today's ocean floor. For example, about 17 million square kilometers of ocean floor is between 0 and 5 million years old.

Graph 1—Area of ocean floor rock versus age

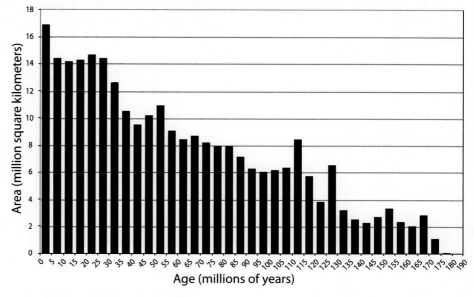

11. How much of the ocean floor is between 75 and 80 million years old?

12. Using Graph 1, summarize the relationship between age and area of the present ocean floor.

Scientists have studied this relationship to try to determine the reason for the difference in the amount of older and younger oceanic rocks. We know that Earth has not grown in size in the past 4.6 billion years. So, if new crust is added in one place by some process, it must be destroyed somewhere else. Earth's systems and processes are connected—if the rate or amount of one process increases or decreases, it causes changes in other processes. We also know that the ocean basins have existed for 3.8 billion years, but the oldest rocks of the ocean floor are a mere 190 million years old. (Older ocean crust has been preserved on some continents.) What explains these observations?

13. Several hypotheses have been proposed to explain this difference in the amount of older and younger oceanic rocks. Consider each of the following hypotheses and describe the information or evidence you would need to support or refute each one.

   a. *"The processes that form oceanic crust have sped up over time, resulting in more young ocean crust."*

   b. *"The total area of all ocean basins has increased over time."*

   c. *"Old oceanic rocks are being destroyed over time."*

   d. *"Old oceanic rocks turn into continental rocks with time."*

   ▶ Turn off the **Ocean Floor Age** and **Continent Age** themes.

Formation of the Atlantic and Pacific Oceans

The ages of ocean floor rocks can be used to determine when individual ocean basins formed and the rate at which they have expanded. Next, you will investigate the ages and expansion rates of the Atlantic and Pacific Ocean basins.

   ▶ Click the QuickLoad button [QL] and select the **Atlantic Basin** extent.
   ▶ Turn on the **Detailed Ocean Floor Age (Ma)** theme.

14. When did the Atlantic basin first begin to open up? How did you determine this?

15. Did the North Atlantic Ocean basin open at the same time as the South Atlantic Ocean basin? Explain. (Hint: Compare the ages of the oldest rocks in each basin.)

Measuring width

To measure the width of the 0-10 million year band (dark red):

- Click on one side of the band.
- Drag across to the other side of the band, parallel to the transform faults along the ridge.
- Read the width (reported as Length) in the status bar above or below the map window.
- Double-click to stop measuring.

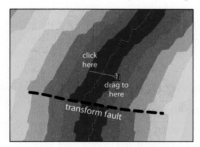

Figure 2. Measuring total width of new sea floor created.

Converting units

To convert from width to rate, divide the width in kilometers by 10 million years:

$$\frac{\text{width (km)}}{10,000,000 \text{ yr}} = \text{rate (km/yr)}$$

To convert the rate from kilometers per year to centimeters to year, multiply by 100,000:

$$\frac{X \text{ (km)}}{1 \text{ yr}} \times \frac{100,000 \text{ cm}}{1 \text{ km}}$$

$$= \text{rate (cm/yr)}$$

▶ Turn on the **Ridges** theme.

▶ Click the QuickLoad button [QL] and select the **North Atlantic** extent.

▶ Use the Measure tool [⊞] to determine the total width of new sea floor that has been created in the North Atlantic over the last 10 million years. (See Figure 2.)

16. What is the total width, in kilometers, of new ocean floor that has been created in the North Atlantic in the last 10 million years?

17. Calculate the rate of ocean floor spreading in the North Atlantic by dividing the width of new ocean floor that you just measured by the time required to create it (10 million years). Convert your result to centimeters per year. (See *Converting units* at bottom left.)

▶ Click the QuickLoad button [QL] and select the **South Atlantic** extent.

▶ Use the Measure tool [⊞] to determine the total width of new sea floor created in the South Atlantic over the last 10 million years.

18. What is the width, in kilometers, of new ocean floor that has been created in the South Atlantic in the last 10 million years? Which part of the Atlantic is spreading at a faster rate—the North or South Atlantic?

19. Using this measurement, calculate the spreading rate of the South Atlantic Ocean basin, in centimeters per year.

▶ Click the Zoom to Full Extent button [🌐] to view the entire map.

Rotate panel

Click the Increment button $\boxed{I}$ to set the rotation increment and / or center coordinates of the map.

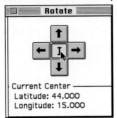

Now look at the Pacific Ocean, off the coast of South America. A disadvantage of flat maps of Earth is that it is not obvious that a feature on the left edge of the map continues on the right edge. Here, the Pacific Ocean appears split into two separate parts. By viewing Earth as a globe, the Pacific appears as it should—a single, continuous ocean basin.

▶ To view the map as a globe, click the Switch Projection button. Click the Increment button $\boxed{I}$ in the Rotate panel, set the center to **-15** Latitude and **-115** Longitude, and click **OK**.

20. How old are the oldest rocks of the Pacific Ocean floor?

▶ Use the Measure tool to determine the greatest width of new ocean floor created in the Pacific Ocean basin over the last 10 million years. **(Caution! See Figure 3.)**

21. How much new ocean floor (total width) has been created at this location in the past 10 million years?

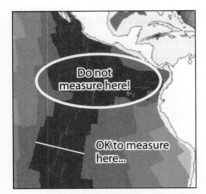

22. Use this measurement to calculate the rate of ocean floor spreading at this location in the Pacific Ocean basin, in centimeters per year.

23. In which ocean basin did you measure the highest spreading rate— the Pacific or the Atlantic?

Figure 3. Caution! Do not measure the width of new ocean floor in the region of the eastern Pacific Ocean shown above. Two ridges (blue lines) intersect in this area, making it difficult to measure either one separately.

▶ Choose **File > Exit** to quit ArcView GIS. Do not save your changes to the project file.

Big ideas and questions about forming ocean basins

Three major ideas presented in this activity are still unresolved:

- The oldest oceanic rocks are 190 million years old, whereas the oldest continental rocks are 3.8 billion (3800 million) years old. Yet we know the oceans existed 3.8 billion years ago. Has old oceanic rock disappeared? Was it converted into something else, like continental rock? Have the oceans been growing at the expense of the continents?

- Ocean floor ages increase laterally with distance away from the youngest rocks. The youngest rocks are typically (though not always) found near the middle of an ocean and the oldest rocks are typically found near the continents. This implies a mechanism whereby ocean crust is made at these central areas, but does not explain how.

- New ocean floor seems to form at different rates in different ocean basins and over time. What is the impact of one ocean basin generating new rock more rapidly than another?

Keep these ideas and questions in mind as you explore the ocean basins in greater detail later in this unit.

Activity 1.3

Ocean origins

How did the oceans form?

Scientists believe that the oceans developed early in Earth's history—at least 3.8 billion years ago. When our planet formed 4.6 billion years ago, heat from compression, nuclear reactions, and collisions with solar system debris caused the early Earth to melt. Molten materials separate, or *differentiate*, into layers based on *density*. In Earth, this process formed concentric layers, beginning with a dense iron-nickel core, surrounded by a thick mantle made of magnesium, iron, silicon, and calcium (Figure 1). The rock with the lowest density rose to the surface to form a thin crust. Together, the crust and rigid part of the upper mantle form the *lithosphere*, with an average thickness of around 100 km. Beneath the lithosphere is the *asthenosphere*, a ductile or plastic region of the upper mantle.

During differentiation, volatile elements trapped within the early Earth rose towards the surface and were vented by volcanoes to form our atmosphere. This early atmosphere consisted mainly of hydrogen and helium, very light elements that were easily lost to space. Around 3.9 billion years ago, as Earth condensed further, a second atmosphere formed, but it contained only traces of free oxygen (O_2). This early atmosphere reflected much of the solar radiation striking Earth, allowing the surface to cool and water vapor to condense into rain. At first, Earth's surface was too hot for liquid water to exist on the surface. Eventually, it cooled enough for water to accumulate, forming the early oceans. Scientists think that this is when the crust began to differentiate into two types—continental and oceanic crust—because the process that forms granite, the most common type of continental rock, requires the presence of water.

The earliest forms of life—blue-green algae—developed in the oceans, where water offered protection from the harmful ultraviolet (UV) radiation that penetrated Earth's early atmosphere. Through the process of photosynthesis, these organisms produced their own food using carbon dioxide, water, and energy from the sun, and released an important by-product—oxygen.

density—the ratio of mass to volume of an object or substance. Mathematically,

$$density = \frac{mass}{volume}$$

plastic—(adj) able to change shape without breaking.

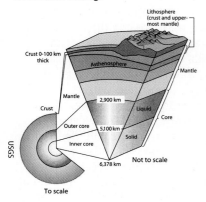

Figure 1. Earth's internal structure is defined by differences in density, composition, and the physical state of the material.

volatile—(adj) easily evaporated.

Why are the ocean basins so much younger than the continents?

To understand the answer to this question, you must examine the processes that form and modify oceanic and continental lithosphere.

Plate tectonics

Earth's outermost layer, the *lithosphere*, is made up of the crust and uppermost part of the mantle. The theory of *plate tectonics* states that the lithosphere is fragmented into twenty or so rigid *plates* that are moving relative to one another (Figure 2). The plates move atop the more mobile, plastic asthenosphere.

Figure 2. Earth's major lithospheric plates.

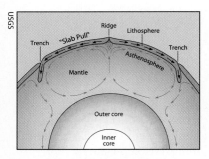

Figure 3. Convection in Earth's mantle transports heat energy to the surface.

Earth's core reaches temperatures over 6,000°C. About half of this heat comes from the decay of naturally-occurring radioactive minerals. The rest is left over from the heat generated when Earth formed. Heat energy travels slowly from Earth's interior to the surface by the processes of *convection* and *conduction*.

In conduction, heat energy moves by collisions between atoms. Convection occurs as rock heats up, expands, becomes less dense, and rises toward the surface (Figure 3). Near the surface it cools, contracts, becomes denser, and slowly sinks deep into the mantle again. Convection cools Earth more efficiently than does conduction. Still, it takes a long time—billions of years—to cool an object the size of Earth. As long as sufficient heat reaches the upper mantle, Earth's lithospheric plates will continue to move.

Types of plate boundaries

There are three types of plate boundaries, classified according to how adjacent plates are moving relative to each other (Figure 4). At *divergent boundaries*, plates move away from each other, forming rift valleys on land and spreading ridges in the oceans. These mid-ocean ridges form the longest mountain belt on Earth, extending over 60,000 km on the ocean floor. At the ridge, molten rock or magma wells up between separating plates and intrudes into the plates, solidifying into new oceanic crust along the ridge crest. Pressure from beneath the ridge, combined with gravity, causes the plates to continually break during earthquakes and spread apart, allowing more magma to rise to the surface, creating more new crust. As the new crust moves away from the ridge, it cools and contracts, becoming denser. Over time, the upper mantle beneath this crust cools, thickening the lithosphere. As the density of the crust and lithosphere increase, they sink deeper into the mantle and the ocean basin deepens. The high ridge at the plate boundary is stationary while the plate itself moves away, revealing its generally flat topography.

At *convergent boundaries*, the plates are moving toward each other (Figure 4). If the two plates are topped with continental crust, they will collide, causing them to pile up and create mountains. If one or both plates

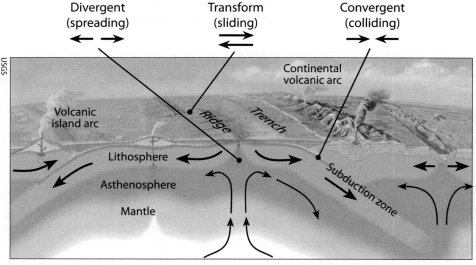

Figure 4. Types of plate boundaries.

are composed of only oceanic lithosphere, one of the oceanic plates will plunge down into the mantle, in a process called *subduction*. Both plates are bent downward along their line of contact, forming a deep trench at the surface along the plate boundary.

The subducting oceanic plate carries with it sediments and water from the surface. Near a depth of around 150 km, the descending plate heats up sufficiently for some of the material to melt. The water and sediments change the chemistry of the magma, forming new minerals. This molten rock rises toward the surface and cools, forming new continental rock.

At *transform boundaries*, plates move past each other (Figure 4). The most famous transform boundary in North America is the San Andreas fault in California. Smaller transform faults accommodate movement along spreading ridges, allowing segments of the ridge to expand independently. These types of boundaries and faults are characterized by frequent earthquakes. Transform boundaries are important in rearranging the plates, but are not directly involved in creating or recycling crustal rock.

The plates have been created, destroyed, and rearranged countless times over Earth's 4.6 billion-year history. New ocean floor forms at mid-ocean ridges and old ocean floor is recycled into the mantle at trenches. Gravity will not allow Earth to expand or shrink, so the net amount of new rock formed must balance the amount being recycled. Since the oldest oceanic crust is about 190 million years old, it is reasonable to say that the entire ocean floor, covering some 70% of Earth's surface, is completely recycled every 190 million years at the current rates of motion.

On the other hand, the low-density continental rock remains on the surface and is not recycled except through weathering and erosion. In fact, continental rocks that formed 3.8 billion years ago can still be found on Earth's surface today, along with continental rock that formed more recently. Thus, plate tectonics and density differences between continental and oceanic rocks explain why the ocean basins are much younger than the continents.

1. If new ocean crust is formed at mid-ocean ridges, where would you expect the oldest ocean floor to be found?

2. At which type of plate boundary does each of the following types of crust form?
   a. Oceanic
   b. Continental

Why are the ocean basins so much lower in elevation than the continents?

Earth's surface has two main features: deep ocean basins and elevated continents. This uneven distribution of topography is no coincidence—it occurs because of gravity and the fact that the continental and oceanic crust are made of different types of rock and have different thicknesses.

The ocean floor is primarily composed of *basalt* (buh-SALT) and *gabbro* (GAB-bro), dark-colored igneous rocks with the same composition and density (~2.9 g/cm³), but which form in different environments. Basalt comes from magma that erupts *onto* the sea floor, while gabbro crystallizes from magma that cools *within* the oceanic crust. Oceanic crust averages around 6 km thick, except at oceanic plateaus where it may reach a thickness of up to 40 km.

The continents are composed primarily of *granite*, a light-colored igneous rock with a density of around 2.7 g/cm³. Continental crust averages about 40 km thick, varying from as little as 20 km to as much as 70 km at high-elevation continental plateaus such as the Tibetan Plateau (the Himalayas).

The existence of deep ocean basins and high-elevation continents is due to the principle of *isostasy* (eye-SAHS-tuh-see), which states that *the total weight of any column of rock and water from Earth's surface to a constant depth is approximately the same as the weight of any other column of equal area* (Figure 5). Differences in the heights of the columns are due to differences in the density and thickness of the materials in the columns. The equilibrium between columns is maintained by the plastic flow of material within Earth's mantle. Though rock in the mantle is generally not hot enough to melt, it is warm enough to flow or change shape very slowly when subjected to a force, much like modeling clay or Silly Putty®.

In the simplest example, isostasy is the principle observed by Archimedes in the third century BC, when he noticed that an object immersed in water displaces a volume of water equal to that of the object. When two columns of different thickness and weight form on Earth's surface, the mantle will slowly flow out from beneath the added load. The region with greater weight will have less mantle beneath it than the region with lesser weight.

If you remove crust or water from the top of the column, mantle material flows in underneath the column to equalize its weight, pushing the column back up slightly. On the other hand, if you pile more crust or water atop the column, mantle material flows away from the bottom of the column to equalize the weight, causing the column to sink back down slightly. The mantle behaves in a manner similar to that of water when an ice cube is placed in a container of water. Water beneath the ice flows outward and upward and the ice sinks downward until the water and ice reach *equilibrium*.

3. Why were there no oceans during the first 800 million years of Earth's history?

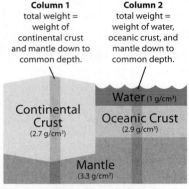

Column 1
total weight = weight of continental crust and mantle down to common depth.

Column 2
total weight = weight of water, oceanic crust, and mantle down to common depth.

Water (1 g/cm³)

Continental Crust (2.7 g/cm³)

Oceanic Crust (2.9 g/cm³)

Mantle (3.3 g/cm³)

Figure 5. The principle of isostasy. Columns with the same area and to the same depth have equal weight.

Exploring isostasy

To explore the principle of isostasy, point your web browser to:

http://atlas.geo.cornell.edu/ education/student/isostasy.html

You can also:

- Open the etoe_unit_1.apr project file.
- Open any view.
- Click the Media Viewer button 🖾.
- Choose **Explore Isostasy.**

equilibrium—state of balance, condition where no overall changes occur in a system.

4. Why do scientists think that the continental rocks could not have formed before the oceans?

5. If a 1 km-thick layer of material were removed from the top of a continent by erosion, would the elevation of the continent decrease by 1 km? Explain.

How do we know what the continents and oceans looked like in the past?

Over the past 200 years, scientists have collected evidence of Earth's changing surface through mapping and monitoring topography; studying earthquakes, volcanoes, and fossils; and measuring the ages of rocks.

As early as the 1600s, cartographers noticed that the coastlines of continents sometimes appeared to match up, even across oceans, like pieces of a puzzle (Figure 6). Then, in the 1800s, scientists discovered that continents separated by vast oceans contained similar landforms, rocks, and fossils. This discovery suggested that today's continents are in very different places than they were in the past (Figure 7). The theory that the continents moved over time to

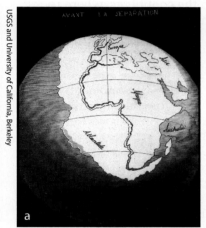

USGS and University of California, Berkeley

Figure 6. In 1858, geographer Antonio Snider-Pellegrini noted how the American and African continents may once have fit together (a), then later separated (b). In the early 1900s, German climatologist Alfred Wegener expanded on this idea in his theory of *continental drift*.

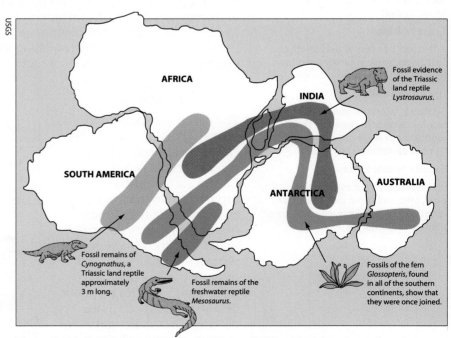

Figure 7. Fossil evidence points to a time around 250 million years ago when the present-day continents were joined in a larger super-continent, now called *Pangea*.

their present locations is called *continental drift*. None of these observations provided an explanation for how entire continents could move over Earth's surface, so the theory was not widely accepted. The key evidence for the mechanism driving plate tectonics came in the mid-1900s, through detailed studies of earthquakes, volcanoes, and the ocean floor.

How old is that rock?

The ages of rocks are determined through a variety of methods. Some techniques allow us to rank the ages of rocks relative to each other, while others allow us to assign numerical ages to individual rocks.

Relative dating

Relative dating methods do not tell us the actual ages of rocks, but they do allow us to determine the chronological order in which rocks formed. Using a series of *stratigraphic principles* (see left), scientists can often reconstruct the history of a sequence of rock layers and geological events. Relative dating techniques are most useful on the continents, where rocks have been uplifted and exposed by weathering and erosion, or can be drilled into relatively easily.

Absolute dating

Absolute dating techniques use laboratory analysis to determine how much time has passed since the rock formed. Most igneous rocks—rocks that form from cooled magma—contain tiny amounts of radioactive elements. These radioactive *parent* elements break down into different *daughter* elements over time at a known rate. By precisely measuring and comparing the amount of parent to daughter elements in a rock, a process called *radiometric dating*, scientists can determine how many years ago the rock formed. Absolute dating techniques are useful for both oceanic and continental rocks, but relatively few oceanic rocks have been dated radiometrically because obtaining good samples is difficult and costly. In practice, absolute and relative dating techniques are often used together to determine approximate ages of rocks.

Paleomagnetism

The ocean floor is composed primarily of an igneous rock called *basalt* that forms as magma erupts at or near Earth's surface and cools rapidly. When the magma cools, some of the mineral grains act like tiny compasses, preserving the direction of Earth's magnetic field. For reasons that are not fully understood, Earth's magnetic field periodically changes polarity. That is, the north and south magnetic poles reverse. This process occurs every

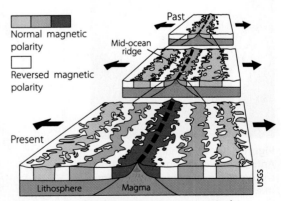

Figure 8. As plates move apart, symmetrical patterns of normal and reversed polarity are preserved in the rocks.

Stratigraphic principles

Original horizontality

Sedimentary layers and lava flows are originally deposited as relatively horizontal sheets.

Lateral continuity

Lava flows and sedimentary layers extend laterally in all directions until they thin to nothing or reach the edge of the basin of deposition.

Superposition

In undisturbed rock layers or lava flows, the oldest is at the bottom and the youngest is at the top.

Inclusions

A piece of rock included in another rock or layer must be older than the rock or layer in which it has been incorporated.

Cross-cutting

A feature that cuts across another feature must be younger than the feature that it cuts.

Unconformities

Surfaces called unconformities represent gaps in geologic time where layers were not deposited or have been removed by erosion.

Faunal (fossil) succession

Plants and animals evolve over time, and each time period can be identified by a unique assemblage of plant and animal fossils.

few hundred thousand years, and each time there is a reversal, minerals in the new rock align differently from those in the older rock next to it. Thus, reversals appear as bands of alternating polarity in the ocean floor. This phenomenon, called *paleomagnetism*, preserves a record of Earth's past magnetic field. At oceanic spreading ridges, the bands appear as symmetrical patterns on either side of the ridge. Like the growth rings of trees, the unique patterns formed by these bands allow scientists to assign ages to the rocks (Figures 8 and 9).

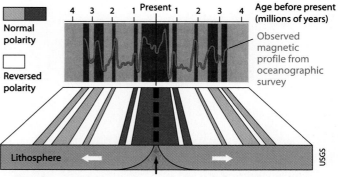

Figure 9. The magnetic polarity profiles are analyzed to determine the age of the ocean floor.

Ships crisscross the oceans, recording these patterns with sensitive *magnetometers*—devices that measure the strength and direction of the magnetic field preserved in oceanic rocks. These data confirm that new ocean floor forms at ridges and moves away from the ridge, providing additional confirmation of plate tectonic theory. In addition, researchers have used paleomagnetic data to measure the rate of sea floor spreading and to reconstruct the history of ocean basin development.

6. Briefly explain the advantages and disadvantages of the following dating techniques for dating ocean floor rock.

   a. radiometric dating

   b. paleomagnetic dating

In this activity, you have seen how heat and gravity drive the processes that are continuously changing Earth's surface. In the next two activities, you will examine the ocean floor at increasing levels of detail to look for features that provide evidence of these processes.

Activity 1.4

Beneath the waves

For most of human history, the ocean's surface has been a forbidding boundary, separating the known from the unknown. Except for the tiny amount of the ocean floor visible in shallow water, people had no idea what lay beneath the waves.

Today, we have the ability to gather detailed information about the age, composition, and other characteristics of the ocean floor that are critical for understanding the processes that shape Earth's surface. In this activity, you will investigate the *bathymetry* (buh-THI-muh-tree) of the ocean basins to more fully understand the features of the ocean floor and the processes that shape them.

bathymetry—measure of water depth in basins such as lakes, oceans, and rivers. *Bathy* comes from a Greek word for *depth*, and *Metry* comes from another Greek word meaning *to measure*.

The five ocean basins

Over 70% of Earth's surface is covered by a single, interconnected body of water. This body of water is somewhat arbitrarily divided into five basins—the Arctic, Atlantic, Indian, Pacific, and Southern Oceans. Before you begin your exploration of ocean bathymetry, you will examine the location, size, and depth of each ocean basin.

▶ Launch ArcView GIS, then locate and open the **etoe_unit_1.apr** project file.

▶ Open the **Ocean Floor Topography** view.

This view shows the ocean basins, each outlined in a different color. If you find the outlines hard to see, turn off the **Countries** theme.

To turn a theme on or off, click its checkbox in the Table of Contents.

To activate a theme, click on its name in the Table of Contents.

▶ Click in each ocean basin using the Identify tool 🛈 and read the name of the basin in the Identify Results window.

1. Label the five global ocean basins on Map 1.

Map 1—Global ocean basins

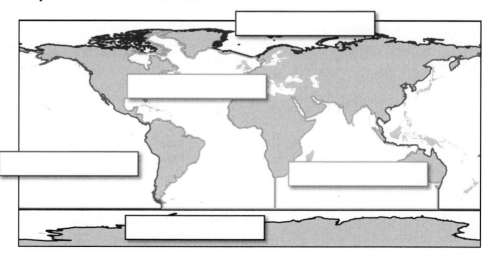

Are there two Pacific Ocean basins?

No. Earth's curved surface has to be "split" somewhere in order to make a flat map. Here, the split was made at 180° longitude, dividing the Pacific Ocean basin into two parts. One part appears on the far right and the other on the far left of the map.

Mapping the ocean floor

Early depth measurements were often labor-intensive and prone to error (Figure 1). Fortunately, today we can map the features of the ocean floor on a global scale using satellites. The satellites measure tiny variations in the height of the ocean surface caused by gravity, which correspond directly to the depth of the sea floor.

Bathymetric profiles

One way to examine ocean depth data is by creating bathymetric profiles that show what the sea floor would look like if you sliced through it and viewed it from the side. These profiles illustrate the shape of the basin and reveal submerged features. Next you will create increasingly detailed profiles to see how our understanding of the ocean basins has improved over the past 150 years.

Like early sailors and explorers, your first profile of the Atlantic Ocean Basin will be based on only a few depth measurements.

▶ Turn on the **Atlantic Crossing** theme. This theme displays the path of a ship crossing the Atlantic Ocean from Florida to Africa.

▶ Click the QuickLoad button and choose the **Atlantic Crossing** extent to zoom in on the ship's path.

2. On Graph 1 sketch what you *think* the profile (side view) of the Atlantic Ocean Basin will look like along the ship's path from North America to Africa. (See *Simple profiles*, lower left for examples.)

Graph 1—Predicted depth profile of Atlantic Ocean Basin

3. Explain your reasoning for the shape of the profile you drew. What determined where you placed the shallow and deep parts of the ocean?

▶ Turn off the **Ocean Basins** theme.
▶ Turn on the **Atlantic Bathymetry (km)** theme.

Figure 1. Manual depth sounding with a weighted cable. It was difficult to obtain accurate depth measurements this way, because there were many sources of error. It was sometimes hard to determine the exact location of the ship, to ensure that the line dropped straight down, and to know when the weight hit the bottom.

Simple profiles

A profile through a bathtub would look something like this:

Whereas a profile of a mountain might look like this:

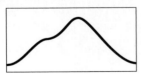

▶ Turn on and activate the **Depth Measurements #1** theme.

The **Depth Measurements #1** theme displays four green points, labeled 1 through 4, at regular intervals along the ship's path. Each point represents a location where the ship stopped to measure and record the ocean depth.

▶ Using the Identify tool 🛈, click on each of the four points along the path. Read the depth at each point from the Identify Results window.

4. Record the depth values for each of the four points in Table 1. Round values to the nearest 0.1 km. (Points 0 and 5 are the coasts of North America and Africa, and are at sea level, where the depth is 0 km.)

Why is the depth negative?

The elevation of a mountain like Mount Everest is based on measuring from sea level to the top of the mountain. In these materials, we will express the depth of a body of water from sea level downward, using negative values. Therefore, we will say the depth of the Mariana Trench is -11,035 meters.

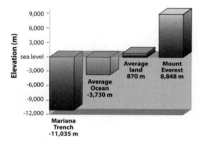

Figure 2. Comparative elevations.

Table 1—Atlantic Ocean Basin Depth Measurements #1

Point	Depth (km)
0	*0*
1	
2	
3	
4	
5	*0*

▶ Close the Identify Results window(s).

5. Plot the depth values from Table 1 on Graph 2. Draw a smooth line through each point beginning at point 0 (sea level) on the North American coast and ending at point 5 (sea level) on the African coast.

Graph 2—Atlantic Ocean Basin depth profile 1

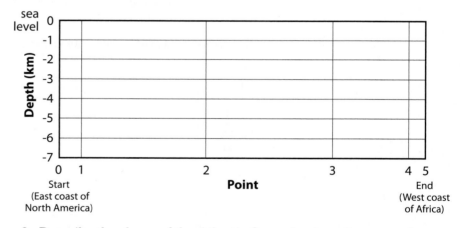

6. Describe the shape of the Atlantic Ocean basin as it appears in Graph 2. How does it differ from your predicted profile (Graph 1)?

Next, you will create a bathymetric profile of the Atlantic Ocean using measurements taken at different points along the ship's path.

▶ Turn off the **Depth Measurements #1** theme and turn on and activate the **Depth Measurements #2** theme.

This theme displays five blue points that represent the locations of the new depth measurements.

▶ Use the Identify tool 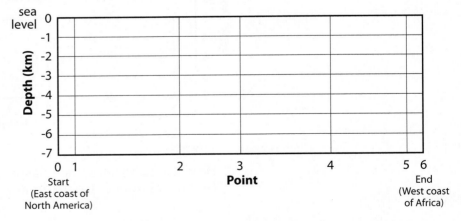 to find the depth at each of the five points along the path.

7. Record the depth values for each of the five points in Table 2. Round values to the nearest 0.1 km. (Points 0 and 6 are the coasts of Florida and Africa, and are at sea level, where the depth is 0 km.)

Table 2—Atlantic Ocean Basin Depth Measurements #2

Point	Depth (km)
0	0
1	
2	
3	
4	
5	
6	0

▶ Close the Identify Results window(s).

8. Plot the depth values from Table 2 on Graph 3 below. Draw a smooth line through each point beginning at point 0 (sea level) on the Florida coast and ending at point 6 (sea level) on the African coast.

Graph 3—Atlantic Ocean Basin depth profile 2

9. Compare Graph 3 to Graph 2. Describe any "new" features that appeared in Graph 3 that are not visible in Graph 2.

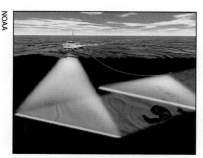

Figure 3. Sonar measures the depth of the water by determining the time it takes the sound waves to travel to the sea floor and back.

Select By Theme button

The Select By Theme button is located on the Button Bar (the top row of buttons).

10. How do the number and selection of measurement locations affect the profile?

▸ Turn off the **Depth Measurements #2** and **Atlantic Crossing** themes.

In the last two profiles you created, you used depth measurements taken at only a few locations. Next, you will create a bathymetric profile of the Atlantic Ocean basin along the ship's path using nearly continuous depth measurements like those generated by sonar (Figure 3).

▸ Click the Select By Theme button 🔲. (Location shown at left.)

▸ Select features from the **Atlantic Bathymetry (km)** theme that **Intersect all the features of** the **Atlantic Crossing** theme and click the **New** button, as shown below.

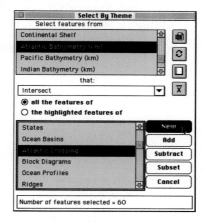

▸ Close the Select by Theme dialog box.

Sixty bathymetry measurements along the Atlantic Crossing path should be highlighted yellow. Next, you will chart these detailed depth measurements.

▸ Activate the **Atlantic Bathymetry (km)** theme.

▸ Click the Open Theme Table button 🖼 to open the **Atlantic Bathymetry (km)** theme table. (Do not click in the table window, or you will lose the features you selected in the previous step.)

▸ Click the Create Chart button 📊.

▸ In the Chart Properties window, choose the **Depth (km)** field and click **Add**. Click **OK** to create the chart.

To make the top of the chart represent sea level, you must modify the chart's vertical scale.

Chart Element Properties tool

The Chart Element Properties tool is located on the Tool Bar (the bottom row of buttons).

▶ Using the Chart Element Properties tool [icon] (see left for location), click on your chart's Y (vertical) axis. This will open the Chart Axis Properties dialog box.

▶ In the Chart Axis Properties dialog box, set the minimum (**Scale min**) to -7 and the maximum (**Scale max**) to 0 as shown below and click **OK**.

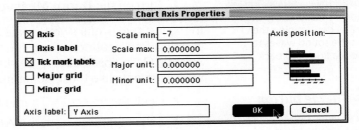

The chart now displays depth measurements along the ship's path across the Atlantic Ocean, creating a bathymetric profile.

11. Using this chart, describe the features that characterize the floor of the Atlantic Ocean basin. (Hint: Click and drag on the corner of your chart to enlarge it. This will help you see the ocean floor in more detail.)

12. How did increasing the number of depth measurements change your view of the ocean floor bathymetry in the Atlantic Ocean basin?

▶ Choose **File > Exit** to quit ArcView. Do not save changes to the project file.

Activity 1.5

Ocean basin features

Early explorers learned (often the hard way) that the ocean floor is *not* smooth like a bathtub. But how does bathymetry vary among ocean basins or within a basin? And how is the shape of the basin related to the age of the rocks or the way the basin formed? To answer these questions, you will begin by looking at the general characteristics of each basin. Then you will examine, in detail, the floors of the ocean basins.

▶ Launch ArcView GIS, then locate and open the **etoe_unit_1.apr** project file.

▶ Open the **Ocean Floor Topography** view.

▶ Using the Identify tool 🛈, click within each ocean basin to obtain its average depth and surface area.

1. Record the average depth and surface area of each ocean basin in Table 1. Then, rank the oceans by area and by depth with 1 = largest or deepest and 5 = smallest or shallowest. Round all values to the nearest 0.1.

Table 1—Global ocean basin statistics

Ocean	Average Depth (km)	Depth Rank	Area (million km²)	Area Rank
Arctic				
Atlantic				
Indian				
Pacific				
Southern				

2. Which ocean basin is the shallowest? How much shallower is this basin than the next shallowest basin?

▶ Close the Identify Results window.

It is clear that the Arctic Ocean is significantly shallower than the other oceans. Next, you will explore why. To do this, you will change the way the depth for each ocean is displayed. Currently, all depths are represented by a single shade of blue. You will load a new legend that classifies depth using different colors. (To speed things up, you will not turn the themes on until *after* you have loaded all of the legends.)

▶ Click the QuickLoad button 🔍, select the **Atlantic Bathymetry** legend, and click **OK**. Repeat this process to load legends for the **Arctic**, **Indian**, **Pacific**, and **Southern Ocean Bathymetry** themes.

▶ Turn on all five of the ocean bathymetry themes (Atlantic, Arctic, Indian, Pacific, and Southern).

Ocean depth is now displayed in shades of blue and purple.

3. List the color(s) and depth range of:

   a. the shallowest part of the ocean

   b. the deepest part of the ocean

▶ Turn on the **Continental Shelf** theme.

This theme shows where continental rocks are submerged beneath ocean waters. This submerged land, called the *continental shelf*, extends as far as several hundred kilometers offshore and to a depth of about 200 meters (Figure 1). Beyond the edge of the shelf, the steep *continental slope* drops away toward the *continental rise* and eventually the deep ocean floor.

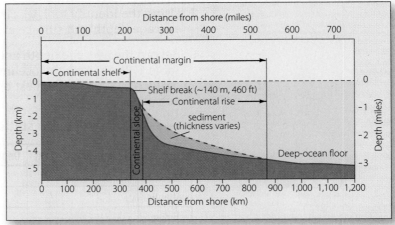

Figure 1. The continental margin.

4. Which ocean has the greatest area of continental shelf?

Ridge flyby animation

To view a simulated "flight" along a ridge, click the Media Viewer button 🔳 and choose **Mid-Ocean Ridge Flyby** from the media list.

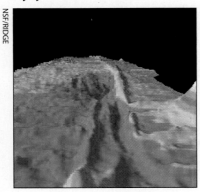

The existence of the continental shelf is not the only reason for shallow ocean floors. Using depth as a guide, locate the submerged mountain range in the Atlantic Ocean basin.

5. Which of the other ocean basins contain similar submerged mountains?

The submerged mountain range you discovered while examining the Atlantic Ocean bathymetry is the Mid-Atlantic Ridge, an example of a spreading ridge. The discovery of the Mid-Atlantic Ridge in the 1930s caused scientists to wonder about the processes that form this type of underwater ridge system and what other types of features may exist on the ocean floor.

▶ Turn on the **Ridges** theme.

The **Ridges** theme displays the locations of mid-ocean ridges.

6. On Map 1, sketch the locations of the oceanic spreading ridges.

Map 1—Oceanic ridges and trenches

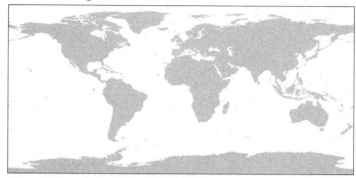

▶ Turn on the **Ocean Floor Age (Ma)** theme.

7. Describe any patterns you observe between the location of the ridges and the age of the ocean rocks.

▶ Turn off the **Ocean Floor Age (Ma)** theme.

Next you will look to see if there is a relationship between the age of the ocean floor and its depth. You will first examine the rocks with ages between 0 and 120 million years.

Graph 1—Average depth versus ocean floor age

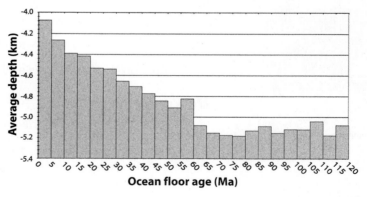

8. Based on Graph 1, describe the relationship between sea floor age and depth.

Hot tips for Hot Links

When you use the Hot Link tool ⚡:

- The theme containing the hot links must be active (the tool on the tool bar will be dark ⚡, not grayed out ⚡.)
- Click the *tip* of the lightning bolt cursor on the feature.

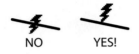

NO   YES!

Trench flythrough animation

To view a simulated "flight" through a trench, click the Media Viewer button 🈁 and choose **Trench Flythrough** from the media list.

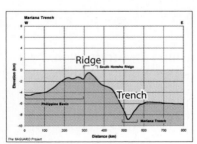

Figure 2. Identifying ridges and trenches in a bathymetric profile.

To better understand the processes that form ridges, look at a block diagram of the Mid-Atlantic Ridge.

▶ Turn on and activate the **Block Diagrams** theme.

▶ Using the Hot Link tool ⚡, click on the profile line that crosses the Mid-Atlantic Ridge to view a block diagram of the ocean floor. (Hint: it is in the same location as the Atlantic Crossing line.)

9. Where is the oceanic lithosphere thicker—near the center of the ridge, or far away from it?

▶ Close the block diagram window.

▶ Turn off the **Block Diagrams** theme.

The spreading ridges are locations where new ocean crust is created. The young, hot, new crust is less dense and rises higher than the surrounding older, colder, more dense rock due to *isostasy*—the same process that allows the continents to rise higher than the ocean floor.

In addition to the continental shelves and mid-ocean ridges, there is another type of feature that reveals how the oceans are created and destroyed. You will explore examples of this type of feature next.

Where does old ocean crust go?

If the deepest parts of the ocean basins are not in the middle of the basins, then where are they?

▶ Examine the legend for the bathymetry themes.

Locate three regions on the map with depths greater than 6 km. They should appear dark blue or purple on the map. These regions are *trenches*, the deepest parts of the oceans. At trenches, oceanic lithosphere is subducted and plunges down into the mantle to be recycled.

10. On Map 1, mark and label (**T**) the locations of three trenches.

▶ Turn off the **Countries** and **Ridges** themes.

Down the trenches

▶ Turn on and activate the **Ocean Profiles** theme.

The **Ocean Profiles** theme displays bathymetric profiles of selected regions of the ocean floor (see Figure 2 for help with interpreting bathymetric profiles).

▶ Use the Hot Link tool ⚡ to click on profiles that cross areas you think contain trenches.

11. Record the name of a trench in the Atlantic Ocean basin. Use the profile graph to estimate its maximum depth and record it below.

Topographic profile movie

To better understand topographic profiles and how they are related to block diagrams, click the Media Viewer button 🔳 and choose **Topographic Profiles** from the media list.

12. Record the name of a trench in the Pacific Ocean basin. Use the profile graph to estimate its maximum depth and record it below.

▶ Turn off the **Ocean Profiles** theme.

The process of creating and destroying the ocean floor results in the opening and closing of ocean basins. The formation of sea floor at ridges often begins by breaking apart a continent. As the ridge develops, so does the ocean floor. Over time, a major ocean basin may form. The formation of the early Atlantic Ocean basin is illustrated in Figure 3.

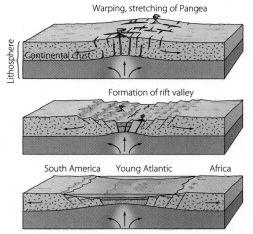

Figure 3. The development of the Atlantic Ocean basin.

You may also view an animation showing how the processes of ridge spreading and subduction work.

▶ Click the Media Viewer button 🔳 and choose the **Ridge Spreading and Subduction** movie from the media list.

What will happen next?

Changes in the shape of ocean basins occur continuously as new sea floor is created at ridges and destroyed at trenches. The tectonic plates can reveal where more crust is being created than destroyed.

▶ Turn on the **Ridges** and **Trenches** themes.

These themes show the locations of spreading ridges and trenches. Each ridge and trench represents a plate boundary.

▶ Turn on the **Relative Plate Motion (mm/yr)** theme.

This theme shows pairs of arrows on opposite sides of plate boundaries. The size of the arrow represents the speed at which the plates are moving relative to each other at that location. Each arrow points in the direction that its plate is moving relative to the neighboring plate.

13. Examine the Atlantic Ocean basin, including any ridges, trenches, and relative plate motion arrows. Would you expect this ocean basin to grow larger or become smaller over time? Explain.

14. Based on surface features and relative plate motions, would you expect the Pacific Ocean basin to grow larger or become smaller over time? Explain.

In this investigation, you have seen that discoveries in the past hundred years have led to a revolution in our understanding of the ocean floor and the processes that shape it. This knowledge has provided important clues to help us interpret Earth's present condition, reconstruct its past, and predict its future.

▶ Choose **File > Exit** to quit ArcView GIS, and do not save changes to the project file.

Activity 1.6

Why are oceans young?

You have examined the ages of both the ocean floor and the continents. Consider some possible explanations, or hypotheses, that might explain why the oceans are generally so much younger than the continents.

1. Read each of the hypotheses presented below and indicate whether the available scientific evidence suggests each explanation could be true, maybe true, or false. For each hypothesis, describe the evidence that supports your choice.

Hypothesis	True	Maybe True	False	Evidence
H1—Earth is only 190 million years old.				
H2—Earth is very old, but the oceans started forming only 190 million years ago.				
H3—The production of ocean crust has increased over the past 190 million years.				
H4—Each year some ocean crust is destroyed, so the amount of ocean crust of a given geologic period decreases over time.				

2. What other hypotheses, both reasonable and far-fetched, can you
   suggest for the young age of the oceans?

3. Based on the evidence you have examined, which hypothesis do you
   feel best explains the young age of the ocean basins. Explain.

Unit 2
Ocean Currents

In this unit, you will...

- *Investigate the forces that drive surface currents in the world's oceans*

- *Identify major ocean gyres and their physical properties—temperature, speed, and direction*

- *Correlate current direction and speed with global winds*

- *Examine ocean salinity and temperature patterns and their relationship to deep-water density currents*

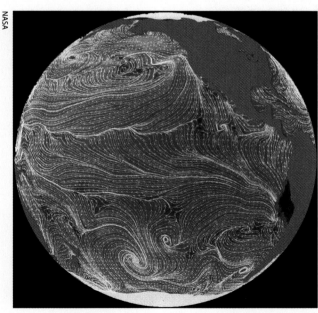

NASA

NASA SEASAT satellite image showing average surface wind speed (colors) and direction (arrows) over the Pacific Ocean.

Activity 2.1

A puzzle at 70°N

"Common sense" tells us that temperatures increase as you move closer to Earth's equator and decrease as you move closer to the poles. If this is true, the pictures below present a strange puzzle (Figure 1). They show two coastal areas at about the same latitude but on opposite sides of the North Atlantic Ocean. Nansen Fjörd, on the left, is on Greenland's eastern coast, while Tromsø, right, lies on the northwestern coast of Norway. These places are at roughly the same latitude, but their climates could hardly be more different.

©2005 L. Micaela Smith. Used with permission.   ©2005 Mari Karlstad, Tromsø Univ. Museum. Used with permission.

Figure 1. A summer day at Nansen Fjörd, on the eastern coast of Greenland (left) and in Tromsø, on the northwestern coast of Norway (right). Both locations are near latitude 70°N.

1. Why do you think the temperatures at the same latitude in Greenland and Norway are so different?

Since the first seafarers began traveling the world's oceans thousands of years ago, we have known about currents—"rivers in the ocean"—that flow over long distances along predictable paths.

In 1855, Matthew Maury wrote about the Gulf Stream current, which flows off the east coast of Florida.

> *"There is a river in the ocean. In the severest droughts it never fails, and in the mightiest floods it never overflows; its banks and its bottom are of cold water, while its current is of warm; the Gulf of Mexico is its fountain, and its mouth is the Arctic Sea. It is the Gulf Stream. There is in the world no other such majestic flow of waters."*
>
> —Matthew Maury, *The Physical Geography of the Sea and Its Meteorology*

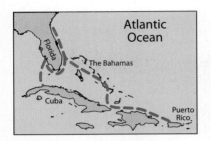

Figure 2. Ponce de León's route

Maury was not the first person to notice the Gulf Stream. In March 1513, the Spanish explorer Juan Ponce de León left the island of Boriquien (Puerto Rico) in search of the island of Bimini and the legendary Fountain of Youth (Figure 2). Instead, he landed on what is now Florida. After sailing northward along Florida's east coast, he turned around and headed south. While sailing in this direction he discovered that even under full sail with a strong breeze at his back, his ship moved backward in the water! His solution was to maneuver his ship closer to shore and out of the current.

Two hundred fifty years later, Benjamin Franklin, then serving as Deputy Postmaster General, received complaints that ships delivering mail between Boston and England took as long as two months to make the return trip back to America. Merchant ships, which were heavier and took a less direct route than the mail ships, were making the trip back from England in just six weeks.

Figure 3. Franklin and Folger's chart of the *gulph stream* current

With help from his cousin Timothy Folger, a whaling captain, Franklin determined that the returning mail ships were sailing against a strong current that ran along the eastern seaboard and across the Atlantic to the British Isles. Whalers knew about the current, whose plankton-rich margins attract whales, and used or avoided the current as needed to speed their travels. Franklin and Folger offered their chart of the *gulph stream* (Figure 3) to the mail ship captains, with the promise of cutting their return time in half, but they were largely ignored.

2. Based on Maury and Franklin's observations, what are some factors that might cause ocean water to flow in currents like the Gulf Stream?

3. Explore the idea of what causes ocean currents by comparing how water behaves in a bathtub or small pond, compared to water in the ocean. In Table 1, make a list of differences between the conditions present or acting on a bathtub of water and an ocean, and explain how those different conditions might cause currents.

Table 1—Comparing bathtub water with ocean water

Condition	Bathtub conditions	Ocean conditions	Why this characteristic might cause currents to form
bottom and surface features			
wind			
volume of water			
salinity			
uniformity of temperature			
Coriolis effect (see note at left)			

Coriolis effect

As a fluid like air or water moves over Earth's surface, the planet rotates under it. Relative to the solid Earth, the flow appears to deflect to the right in the Northern Hemisphere and to the left in the Southern Hemisphere. In this way, the Coriolis effect influences the rotation of large-scale weather and ocean current systems. The Coriolis effect does not influence water in sinks or toilets because the distances involved are very small.

To learn more about the Coriolis effect, point your web browser to:

http://ww2010.atmos.uiuc.edu/ (Gh)/guides/mtr/fw/crls.rxml

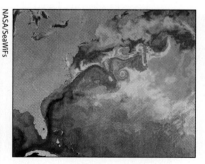

NASA/SeaWiFs

Figure 4. Satellite image of sea surface temperatures associated with the Gulf Stream off the east coast of North America. Reds and oranges represent warm water, greens and blues cooler water. The warmest water appears dark brown or almost black in this image.

4. Which of the conditions above do you think are the most important in the formation of ocean currents? Explain.

Early nautical charts depicting the Gulf Stream current were useful, but were not entirely accurate. They often assumed that the Gulf Stream began in the Gulf of Mexico when, in fact, it flows westward from the equatorial Atlantic Ocean, turns northward and flows along the East Coast from Florida to the Saint Lawrence Seaway, and then across the Atlantic toward Great Britain (Figure 4).

5. Maury and Franklin both described the Gulf Stream as a warm surface current—that is, its water is warmer than the surrounding ocean. Do you think the ocean also has cold surface currents? Explain your reasoning.

Despite Maury's assertion that "There is in the world no other such majestic flow of waters," the Gulf Stream is not unique. Surface currents have existed in the world's ocean basins throughout Earth's history, and have influenced life on our planet in important ways.

6. Describe four ways that surface currents might affect you (or another person), either at sea or on land.

   a.

   b.

   c.

   d.

7. Recall the puzzle posed in Question 1 about the extreme climate differences between the coasts of Greenland and Norway. The map below shows the location of the Gulf Stream current. On the map, draw the locations of other currents that you think could "solve" this puzzle. Label each current as warm or cold.

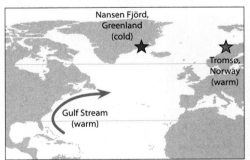

In this unit, you will investigate the forces that drive surface currents and how these currents influence ocean processes and life on Earth.

Activity 2.2

Oceans in motion

The oceans are not stagnant, motionless bodies of water. Entire ocean basins gradually change shape, size, and location over millions of years. The oceans are continually active in many other ways as well, across a variety of time scales. Waves rush in and out on an ocean beach within seconds, and sea level rises and falls with the daily tides. Offshore, the movement of the ocean is equally pronounced, with large volumes of water flowing in tremendous currents. Currents can be thought of as vast rivers without banks that transport immense volumes of water around the globe. In this activity, you will explore the characteristics of these surface currents.

How deep are surface currents?

Surface currents typically extend to depths of less than 400 meters.

Surface currents

To understand how surface currents form, you will begin this investigation by exploring where they are found and the directions in which they flow.

▶ Launch ArcView GIS, then locate and open the **etoe_unit_2.apr** project file.

▶ Open the **Surface Currents** view.

The **Surface Currents** theme shows the approximate locations and extents of surface currents in each ocean basin. Different colors represent individual currents, and white represents areas without significant currents.

1. In addition to the Arctic Ocean, there are other large areas in the oceans where surface currents are absent. Where, within the ocean basins, are these areas located?

To turn a theme on or off, click its checkbox in the Table of Contents.

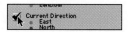

To activate a theme, click on its name in the Table of Contents.

Figure 1. The main, or *cardinal*, directions north, east, south, and west.

▶ Turn off the **Surface Currents** theme. Hide the currents legend (the names and colors of the currents) in the Table of Contents by activating the **Surface Currents** theme and choosing **Theme ▶ Hide | Show Legend**.

Currents in ocean basins

Next, you will search for patterns in the general movement of water within ocean basins.

▶ Turn on and activate the **Current Direction** theme.

In this investigation, current direction is defined as the direction the water is flowing *toward*. Thus, the direction of a current flowing from east to west will be described simply as *west*. The **Current Direction** theme shows the average flow direction of the ocean's surface layer, in one of four cardinal directions (Figure 1): north (N), south (S), east (E), or west (W).

2. Examine the direction of currents in each ocean basin, then draw lines with arrows indicating the direction of motion in each hemisphere and each basin on Map 1. Remember, there is only one Pacific Ocean basin; but in this map projection, half of the basin appears on each side of the map.

Map 1—Generalized ocean currents

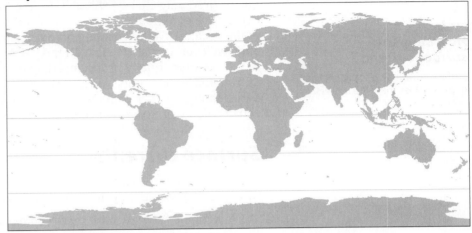

The large, roughly circular paths taken by currents as they flow around the edges of ocean basins are called *gyres* (JYE-urz).

3. Examine Map 1 and compare the circulation of gyres in the Northern and Southern Hemisphere. Complete the table below using **CW** (clockwise), **CCW** (counterclockwise), or **B** (both).

Hemisphere	Circulation
Northern	
Southern	

▶ Turn off the **Current Direction** theme.

A closer look at ocean gyres

The four currents that form a gyre flow in a closed circuit around the outer edge of an ocean basin (Figure 2). The two currents traveling east and west across the basin are called *transverse currents*, and the two flowing north or south near or along the edges of continents are called *boundary currents*. Next, you will examine the unique temperature and speed characteristics of currents within these gyres.

Figure 2. Four types of boundary currents of the South Atlantic gyre.

Temperature patterns

In this section, you will characterize global ocean temperatures and investigate how surface currents may influence them.

▸ Turn on and activate the **Water Temperature (C)** theme.

This theme displays the average annual surface temperature of the world's oceans. Dark red represents warmer temperatures and dark blue represents cooler temperatures.

4. Which latitude bands in each hemisphere contain the warmest surface waters? Which contain the coldest surface waters?

　a. warmest

　b. coldest

5. Given the underlying principle that heat flows from warmer to cooler areas and knowing the direction that the gyres flow in each hemisphere, predict which boundary current—the eastern or western boundary current—will be warmer in each hemisphere. (Refer to Figure 2.)

　a. Northern Hemisphere: western / eastern (circle one)

　b. Southern Hemisphere: western / eastern (circle one)

To check your prediction, calculate the average temperature of the eastern boundary currents in the Northern and Southern Hemispheres. The western boundary currents have already been done for you.

Return here to repeat this procedure.

▸ Click the Query button and query the **Water Temperature (C)** theme for **([Type] = "eastern boundary") and ([Hemisphere] = "Northern")** exactly as shown below.

▸ Set the **Found Features** option to **Highlight**, then click the **New** button.

Liquid water at -1°C?

The **Water Temperature (C)** theme shows water as cold as -1°C. *Pure* water freezes at 0°C, but dissolved salts in ocean water lower its freezing point. The average freezing point of seawater is about -2°C.

QuickLoad Query

If you have difficulty entering the query statement correctly:

• Click the **QuickLoad Query button** and load the **N Hemisphere Eastern Boundary** query.
• Select the **Highlight** option.
• Click **New.**

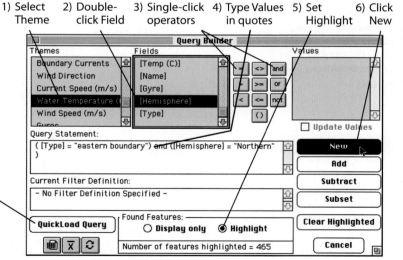

▸ Close the Query Builder window.

The eastern boundary currents in the Northern Hemisphere should be highlighted yellow.

Next, you will calculate the average temperature of these currents.

▶ Click the Statistics button ⎡x̄⎤ and calculate statistics for **the selected features of** the **Water Temperature (C)** theme, using the **Temp (C)** field. Use the **Basic** output option and click **OK**.

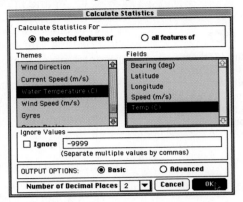

6. Record the average (**Mean**) temperature of the eastern boundary currents in Table 1. Round values to the nearest 0.1°C.

Rounding

To learn more about rounding, see the section on **Rounding** in the Introduction to this guide.

Table 1—Average temperature of boundary currents by hemisphere

Boundary Currents	Average Temperature (°C)	
	Northern Hemisphere	Southern Hemisphere
western	22.9	20.9
eastern		

▶ Close the Statistics window.

▶ Repeat this process to find the mean temperature of the **"Southern"** Hemisphere **"eastern boundary"** currents and complete Table 1.

QuickLoad Query

If you have difficulty entering the query statements correctly:

- Click the **QuickLoad Query button** and load the **S Hemisphere Eastern Boundary** query.
- Select the **Highlight** option.
- Click **New.**

7. How do these results compare with your predictions about the temperatures of boundary currents in question 5? Explain the patterns you see.

8. Use the temperature data in Table 1 and the direction of the currents you recorded on Map 1 to complete these statements about how boundary currents redistribute heat energy within ocean basins. Refer to Figure 2 to review the four types of boundary currents.

   a. Western boundary currents transport (circle one) warm / cold water to (circle one) tropical / polar regions.

   b. Eastern boundary currents transport (circle one) warm / cold water to (circle one) tropical / polar regions.

▶ Close the Statistics and Query Builder windows.

▶ Turn off the **Water Temperature (C)** theme.

Speed patterns

Just as we can predict ocean temperatures based on where a current originates, the location of a current within a gyre also provides clues about the current's speed. Looking at the speed of the current may also help us answer the question "What drives the currents?"

▶ Turn on and activate the **Current Speed (m/s)** theme.

To turn a theme on or off, click its checkbox in the Table of Contents.

To activate a theme, click on its name in the Table of Contents.

This theme displays the average annual speed of water circulating in the world's oceans. Dark red represents higher speed (faster currents), whereas pink represents lower speed (slower currents). Examine the map and look for patterns in the current speed.

9. Which of the four types of boundary currents appear to be moving the fastest in each hemisphere?

Next, you will examine the speed of boundary and transverse currents in gyres by summarizing the speed data based on the type of current.

▶ Click the Summarize button **Σ** and summarize the **Current Speed (m/s)** theme based on the **Type** field. In the table, include the **Speed (m/s)** field summarized by **Average** and click the **Add** button. The window should appear as shown below. Select the **All Values** option and click **OK** to calculate and display the summary table. **(Be patient—summary tables may take a while to process.)**

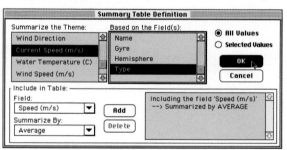

10. Use the summary table to complete Table 2. Round values to the nearest 0.01 m/s. (The order of the current types in the summary table does not match the order in Table 2, so be sure to record the results in the correct rows of the table.)

Table 2—Average speeds of boundary current types

Boundary current type	Average speed (m/s)
Northern transverse	
Southern transverse	
Eastern boundary	
Western boundary	

▶ Close the summary table window.

In the next section, you will explore why these currents flow at different speeds.

Where the wind blows

Global winds are general, consistent patterns of air movement driven by the sun's heat energy and Earth's rotation. Next, you will examine these circulation patterns and compare them to ocean surface current patterns.

Wind and current directions

Normally, currents are labeled according to the direction they are flowing *toward*, whereas winds are labeled according to the direction they are blowing *from*. To avoid confusion, winds and currents are both labeled according to the direction they are moving *toward* in this activity. Thus, a wind or current flowing from south to north is designated **N** or **north**.

Major ocean currents may also be named for their geographic location. For example, the Benguela Current is named after the port city of Benguela, on the coast of western Angola, Africa.

▶ Turn off the **Current Speed (m/s)** and **Ocean Labels** themes and turn on and activate the **Wind Direction** theme.

11. Examine the predominant wind direction within each latitude band. Summarize your observations in Table 3. If there is not a clear overall movement of the wind in one direction, record the wind direction as *mixed*.

Table 3—Wind direction by latitude

Latitude Band	Predominant wind direction
60-90° N	
30-60° N	
0-30° N	
0-30° S	
30-60° S	
60-90° S	

12. How does the wind direction change between 30-60° N latitudes as the winds approach the western edge of both North America and Northern Africa?

13. Does the same change in wind direction occur between 30–60° S as the winds approach the western edge of South America, Southern Africa, and Southern Australia? If not, how does the pattern differ?

14. Based on the currents you drew on Map 1, how do the ocean current directions compare to the predominant wind—the same, opposite, or at some other angle to the wind direction? Explain your answer.

15. Using your knowledge of winds and the location of large land masses, explain why you think transverse currents flow faster than boundary currents.

Surface currents and winds

Surface currents appear to be related to prevailing winds, but do the winds and currents move in exactly the same direction? Next, you will examine four currents to compare the wind and the current direction.

▶ Turn on and activate the **Selected Currents** theme.

The **Selected Currents** theme outlines large segments of four major surface currents: the California, Benguela (ben-GWAY-luh), North Equatorial and South Indian currents. You will be gathering and recording data about these currents in Table 4 below. Data for the Benguela and North Equatorial currents have been entered in the table for you.

16. For the California and South Indian current regions, visually estimate the predominant wind direction and record your answer in the **Direction - Wind** column of Table 4. (Note: It may help to turn the **Selected Currents** theme on and off.)

▶ Turn off the **Wind Direction** theme and turn on the **Current Direction** theme.

The **Current Direction** theme shows the average direction of oceanic surface currents.

17. Visually estimate the predominant current direction for these two regions and record it in the **Direction - Current** column of Table 4. Again, it may help to turn the **Selected Currents** theme on and off.

Table 4—Direction of global winds and surface currents

Surface Current	Direction		Bearing		Wind-current offset direction	Hemi-sphere
	Wind	Current	Wind	Current		
Benguela	N	N	344	331	CCW	S
California						
North Equatorial	W	W	237	275	CW	N
South Indian						

Note: At this point, only the **Direction** section of Table 4 will be completely filled in for all four currents. You will complete the rest of the table later.

18. Compare the general direction of the currents to the direction of the winds in Table 4.

▶ Turn off the **Current Direction** theme and turn on the **Wind Direction** theme.

Next, you will compare your visual observation with each current's average direction (bearing) in degrees.

▶ Click the Query button  and query the **Selected Currents** theme for **[Name] = "California"**.

Return here to repeat this procedure.

▶ Set the **Found Features** option to **Highlight**, click the **New** button, then close the Query Builder window.

QuickLoad Query

If you have difficulty entering the query statement correctly:

- Click the **QuickLoad Query button** and load the **California Current** query.
- Select the **Highlight** option.
- Click **New.**

The outlined region of the California Current should now be highlighted yellow. Next, you will perform a Select by Theme operation to determine the bearing of the wind over the California Current.

▶ Click the Select By Theme button 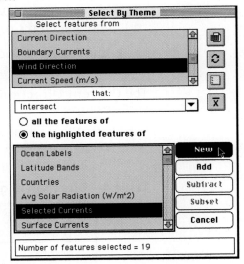 (location shown at left) and select the features from the **Wind Direction** theme that **intersect the highlighted features of** the **Selected Currents** theme as shown below.

Select By Theme button

The Select By Theme button is located on the Button Bar (the top row of buttons).

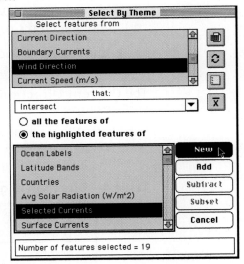

▶ Click the **New** button but do not close the Select By Theme window.

▶ Click the Statistics button ⟨x̄⟩ in the Select By Theme window and calculate statistics for **the selected features of** the **Wind Direction** theme using the **Bearing (deg)** field. Select the **Basic** output option and click **OK**.

19. Round the average (**Mean**) wind bearing to the nearest degree and record it in the **Bearing - Wind** column of Table 4.

▶ Close the Statistics window.

Now you will determine the average direction (bearing) of the California Current in the outlined region.

▶ In the Select By Theme window, select the features from the **Current Direction** theme that **Intersect the highlighted features of** the **Selected Currents** theme.

▶ Click the **New** button but do not close the Select By Theme window.

▶ Click the Statistics button $\boxed{\bar{x}}$ in the Select By Theme window and calculate statistics for **the selected features of** the **Current Direction** theme using the **Bearing (deg)** field. Select the **Basic** output option and click **OK**.

20. Record the average (**Mean**) current bearing in degrees in the **Bearing - Current** column of Table 4.

▶ Close the Statistics window.

▶ Repeat the query and select by theme operations for the "**South Indian**" Current and record the wind and current bearings in Table 4.

21. On Graph 1, plot each surface current from Table 4. (The Benguela Current has been done for you.)

a. Draw a solid line to indicate the average wind bearing.

b. Draw a dashed line to indicate the average current bearing.

c. Draw an arrow from the wind bearing to the current bearing. This arrow represents the wind-current offset.

d. Label the current.

QuickLoad Query

If you have difficulty entering the query statement correctly:

• Click the **QuickLoad Query** button and load the **South Indian Current** query.
• Select the **Highlight** option.
• Click **New.**

Graph 1—Winds and currents of major surface currents

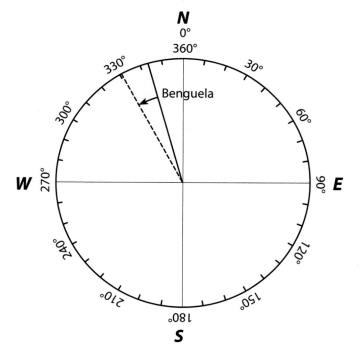

22. Based on the arrows you drew on Graph 1 from each wind to its related current, record whether the wind-current offset for each current is counter-clockwise (CCW) or clockwise (CW) in the **Wind-current offset direction** column of Table 4.

23. Record the hemisphere where each current is located in Table 4. (Northern Hemisphere = N and Southern Hemisphere = S).

24. Using the data you recorded in Table 4, describe any patterns you see between the hemisphere and the wind-current offset direction.

    a. Northern Hemisphere

    b. Southern Hemisphere

25. What do you think causes the small difference in direction between a wind and its associated current? Do you think the wind is driving the current or is the current driving the wind? Explain.

This pattern of offset between the direction of the currents and their associated winds is called *Ekman transport*. This phenomenon is caused by the Coriolis effect and friction between successively deeper layers of water. You will learn more about Ekman transport in the next activity.

▶ Choose **File > Exit** to quit ArcView. Do not save changes to the project file.

Activity 2.3

Figure 1. Global winds (red) and their corresponding surface currents (blue) in the North Atlantic Ocean.

density—mass per unit volume of a substance or object.

$$\text{density (kg/m}^3) = \frac{\text{mass (kg)}}{\text{volume (m}^3)}$$

Current basics

Ocean waters are continuously moving, circling the ocean basins in powerful currents hundreds of kilometers wide, and in swirls and eddies as small as a centimeter across. The primary forces driving the large-scale motions are the sun's energy and Earth's rotation. Energy from the sun warms Earth's surface and atmosphere, generating winds that initiate the horizontal movement of surface water (Figure 1). Vertical movement between the surface and the ocean depths is tied to variations in temperature and salinity, which together alter the *density* of sea water and trigger sinking or rising of water masses. Together, the horizontal and vertical motions of water link the world's oceans in a complex system of surface and subsurface currents often referred to as the *Global Conveyor Belt* (Figure 2). This circulation system plays a vital role in transporting and distributing heat, nutrients, and dissolved gases that support life around the globe.

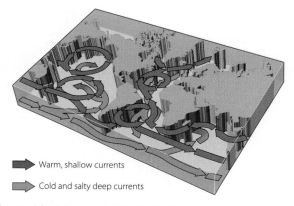

➡ Warm, shallow currents

➡ Cold and salty deep currents

Figure 2. A highly simplified diagram of the *Global Conveyor Belt*.

Structure of the ocean waters

The oceans contain numerous *water masses,* which can be distinguished by their physical and chemical characteristics such as salinity, temperature, and density. The density of seawater depends on its temperature and salinity, as well as the amount of pressure exerted on it. Water expands as it warms, increasing its volume and decreasing its density. As water cools, its volume decreases and its density increases. Salinity, the amount of dissolved solids (like salts) in the water, alters density because the dissolved solids increase the mass of the water without increasing its volume. So, as salinity increases, the density of the water increases. Finally, when the pressure exerted on water increases, its density also increases.

Changing density

The density of water changes as its temperature or salinity (or both) change.

- If the temperature decreases and/or the salinity increases, the water becomes more dense.
- If the temperature increases and/or the salinity decreases, the water becomes less dense.

1. Rank the following types of ocean water from highest density (1) to lowest density (3).

    a. warm, salty water _____

    b. cold, salty water _____

    c. warm, fresh water _____

The characteristics of a water mass typically develop at the ocean surface due to interactions with the atmosphere. Evaporation can increase salinity as fresh water is removed from the ocean and the salts are left behind. Precipitation has the opposite effect, decreasing salinity levels as fresh water is added to the ocean. Processes like *photosynthesis* and the exchange of energy and matter between the ocean surface and the atmosphere can affect the amounts of oxygen and other dissolved gases in the water.

photosynthesis - Process by which organisms convert sunlight and carbon dioxide to carbohydrates (food) and oxygen (O_2).

Thermocline

The thermocline is a layer of the ocean in which the temperature decreases rapidly with depth. Above the thermocline, the temperature is fairly uniform due to the mixing processes of currents and wave action. In the deep ocean below the thermocline, the temperature is cold and stable.

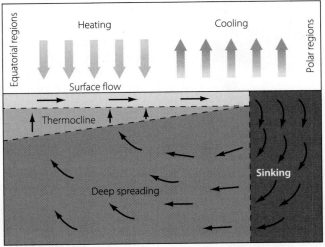

Figure 3. Schematic cross-section of ocean from equator to pole.

In addition, water temperature (and thus density) changes rapidly as surface currents transport water masses from the equator to the poles and vice versa. Although the sun's energy is very efficient at warming the upper 100 meters of the ocean, very little solar energy penetrates to deeper waters. Therefore, water temperature decreases rapidly between 100 and 800 meters depth. This region of decreasing temperature is called the *thermocline*, and marks the boundary between surface water circulation and deep water circulation (Figures 3 and 4).

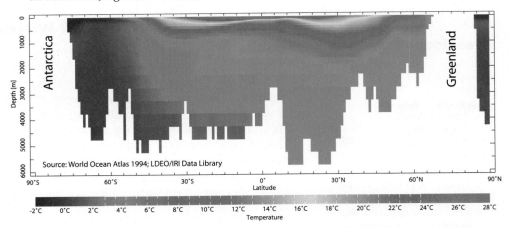

Figure 4. South-north temperature profile of the Atlantic Ocean basin at 32.5° W longitude. White represents the ocean floor and continents.

2. The water temperature at the base of the thermocline is around 5°C. Using this information, sketch and label the approximate location of the base of the thermocline on Figure 4.

A similar zone, in which salinity changes rapidly with depth, is called the *halocline* (Figure 5). However, the halocline is not as well defined as the thermocline and in some places does not exist.

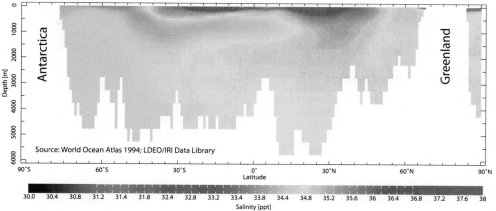

Figure 5. South-north salinity profile of the Atlantic Ocean at 32.5° W longitude. White represents the ocean floor and continents.

Once formed, water masses tend to retain their original characteristics because they mix very slowly with the surrounding water—except in places where the thermocline is very weak. Their distinctive characteristics make it possible to identify their place of origin and track their movements. In fact, it is by tracking differences in the physical properties of water masses that scientists have been able to begin mapping the Global Conveyor Belt.

Wind-driven currents

Winds are created by the uneven heating of Earth's surface by the sun, due primarily to Earth's nearly spherical shape (Figure 6). Surface temperature variations create temperature and pressure differences in the layer of air near the surface. To equalize these differences, air moves from regions of high pressure to regions of low pressure, creating wind.

Spreading light

When the sun is directly overhead at the equator, the same amount of sunlight that falls on one square meter at the equator would be spread over two square meters in Anchorage, Alaska.

In the Tropics, the sun's rays are nearly perpendicular to Earth's surface, producing maximum heating.

Near the poles, Earth's curvature causes the energy to spread over a greater area, producing less surface heating.

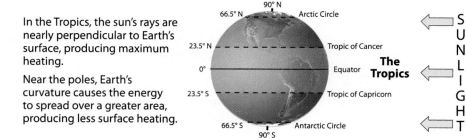

Figure 6. Variation in solar heating with latitude.

Low pressure belts form where warm air rises, near the equator and around 60° latitude (Figure 7); high pressure belts are found where cool air sinks, near the poles and around 30° latitude. Air moving from high pressure toward low pressure creates six global wind belts encircling Earth. These belts shift slightly north and south with the seasons, but they are otherwise permanent features.

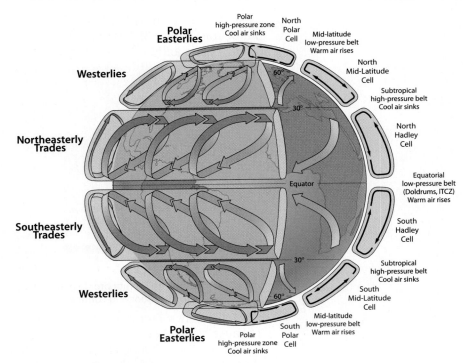

Traditional wind names

The global wind belts in Figure 7 are named, by tradition, according to the direction they are *blowing from*. In these materials we name both winds and ocean currents according to the direction they are *flowing toward*. For example, in the Northern Hemisphere, we would describe the direction of the Westerlies as *northeast* (or *NE*).

Figure 7. Global wind belts.

Strong prevailing winds and warmth from the sun produce ocean surface currents that extend to depths ranging from 45–400 meters under typical conditions. This surface layer of currents is called the *Ekman layer,* or the *wind-blown layer.*

The Coriolis effect and Ekman transport

Over short distances, winds and the ocean surface currents they generate follow straight paths, but over greater distances they curve due to Earth's rotation. This phenomenon is called the *Coriolis effect.* In the Northern Hemisphere, the Coriolis effect causes winds and ocean currents to veer to the right; in the Southern Hemisphere, the winds and ocean currents curve to the left.

It is useful to think of the Ekman layer as containing many thinner layers of water flowing over one another (Figure 8). The wind pushes the topmost layer along and the Coriolis effect deflects it 15–45° *to its right* in the Northern Hemisphere and 15–45° *to its left* in the Southern Hemisphere. As this motion is transferred downward from one layer to the next within the Ekman layer, friction causes additional slowing and deflection. At the bottom of the Ekman layer, it is theoretically possible for water to actually flow in a direction opposite to the surface current, but this has never been observed. The overall motion of the Ekman layer, referred to as *Ekman transport,* is at an angle of about 90° to the wind direction.

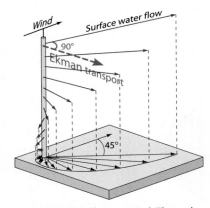

Figure 8. The Ekman spiral. The red arrow represents the net effect, called Ekman transport. (Clockwise Northern Hemisphere deflection shown here—Southern Hemisphere deflection is counterclockwise.)

Note: The water does *not* spiral downward like a whirlpool!

3. If the arrow below represents the prevailing winds somewhere over the ocean in the Southern Hemisphere:

   a. Draw a solid arrow to show the probable direction of the ocean surface current.

   b. Draw a dashed arrow to show the probable direction of Ekman transport.

Wind-driven upwelling and downwelling

In nearshore environments, it is common to have winds blowing parallel to shore over the ocean (Figure 9). Ekman transport moves surface water offshore and pulls deep, cold, nutrient-rich water to the surface. This process, known as *wind-driven upwelling,* is restricted mainly to the west coast of continents, and is responsible for the high productivity of nearshore waters.

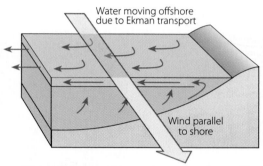

Figure 9. Factors that produce coastal upwelling.

Upwelling occurs in the open ocean near the equator in a similar manner (Figure 10). On both sides of the equator, surface currents moving westward are deflected slightly poleward and are replaced by nutrient-rich, cold water from great depths.

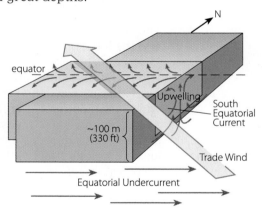

Figure 10. Factors that produce equatorial upwelling.

The mechanical action of wind on the currents promotes mixing of the Ekman layer, which tends to deepen the thermocline and promote the upwelling of nutrients. The thermocline, which separates less dense, warm surface water from the more dense, cold water below, is most pronounced at low latitudes and prevents nutrient rich deep waters from rising to the surface. In contrast, upwelling occurs more readily in high-latitude regions near the poles. These regions receive little sunlight and are not warmed by solar energy. Without a distinct thermocline, upwelling easily brings nutrients toward the surface and promotes mixing.

Surface currents in ocean basins

Gyres play a major role in redistributing the sun's heat energy around the globe. Each gyre consists of four interconnected, yet distinct currents (Figure 11). A pair of *boundary currents* flows north or south, parallel to the bordering landmasses. Western boundary currents carry warm equatorial water poleward, while eastern boundary currents carry cooler temperate and polar water toward the equator. These currents interact with the air near the surface to moderate the climate of coastal regions. Within a gyre, boundary currents are connected by *transverse currents*. Transverse currents move east or west across the gyre's northern and southern edges.

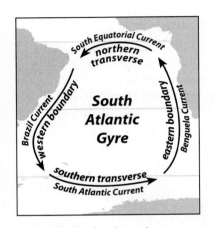

Figure 11. The four boundary currents that form a gyre.

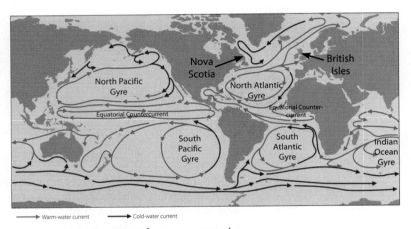

Figure 12. Major ocean surface currents and gyres.

The speed of a current within a gyre is related to the prevailing winds and the location of landmasses. Western boundary currents are narrow but move huge masses of water quickly as the westward-blowing trade winds push water against the eastern edges of continental landmasses. The Coriolis effect and resulting Ekman transport occurring at 90° from the wind direction further enhance the speed of western boundary currents, a phenomenon called *western intensification*. Although most of the water at the equator moves westward and poleward, the low-intensity winds and lack of Coriolis effect at the equator allow for some of the water to flow eastward in equatorial countercurrents (Figure 12).

Figure 13 shows the currents of the North Pacific Gyre. Use what you have learned about surface currents to answer the following questions.

Kuroshio = koor-OH-shee-oh

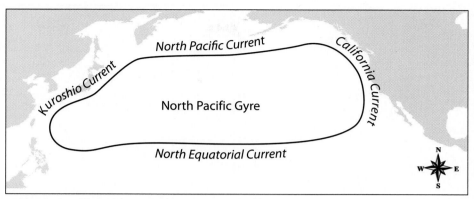

Figure 13. The North Pacific Gyre.

4. Draw arrows on the map to show the direction of each of the four currents labeled.

5. Complete Table 1 with information about the surface currents of the North Pacific Gyre. For **Heat exchange type**, indicate whether the current is gaining heat (warming) or losing heat (cooling) as it flows.

Table 1—Boundary currents of the North Pacific Gyre

Type of current	Name	Heat exchange type (cooling / warming)
Eastern Boundary		
Western Boundary		
Northern Transverse		
Southern Transverse		

Density-driven currents

In addition to the wind-driven horizontal surface currents described above, ocean circulation has a vertical component that is driven by differences in water density. When surface water cools or becomes more saline due to evaporation or other processes, its density increases and it sinks either to the bottom of the ocean or to a depth where its density equals that of the surrounding water. This density-driven circulation pattern is referred to as *thermohaline circulation*, and the currents it produces are called *density currents*. The cold water eventually returns to the surface to be reheated and returned to the poles by surface currents, or to mix with other water masses and return to the depths. Thermohaline currents move very slowly—about 1 centimeter per second—10 to 20 times slower than surface currents.

Figure 14 shows the relationship between temperature, salinity, and density of ocean water. The dashed lines are lines of constant density.

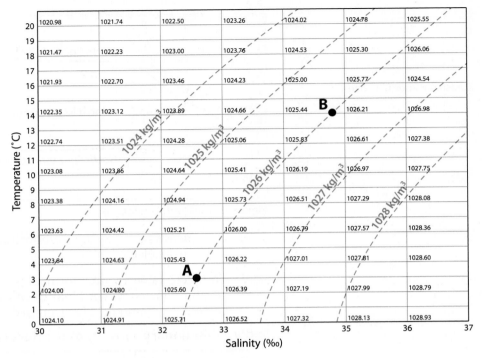

Figure 14. Ocean water Temperature-Salinity-Density chart.

Salinity

The average salinity of sea water is 34.7 ppt or *parts per thousand* (also symbolized ‰). That means that a liter of ocean water (a little more than a quart) contains 34.7 grams (~2.5 tablespoons) of various salts.

To learn more about the composition of seawater, click the Media Viewer button [🖼] and choose **Seawater**.

6. What happens to the density of the water as the temperature decreases? (Follow one of the vertical lines of constant salinity downward, and note what happens to the density values.)

7. What happens to the density of ocean water as the salinity increases? (Follow one of the horizontal lines of constant temperature from left to right and note what happens to the density values.)

Deep currents are generated by relatively small density variations. In fact, the density of seawater must be determined to several decimal places to detect significant differences. The points labeled A and B on Figure 14 represent the salinity and temperature values for two water masses.

8. Use Figure 14 to determine the temperature, salinity, and density of water masses A and B and record them in Table 2.

Table 2—Mixing of water masses A and B

Point	Temperature (°C)	Salinity (ppt)	Density (kg/m³)
A			
B			
C			

When two water masses of the same density meet, they tend to mix. The temperature and salinity of the new water mass lie somewhere between those of the two original water masses. Imagine mixing equal parts of the two water masses. The temperature and salinity of the new water mass would lie at the midpoint of a straight line connecting point A to point B.

9. Draw a straight line connecting points A and B on Figure 14. Plot the midpoint of the line and label it C.

10. Would the density of the new water mass C be higher or lower than the densities of the two original water masses, A and B?

11. Record the temperature and salinity of point C in Table 2. Use the curved equal-density lines to estimate the density of water mass C and record it in Table 2.

12. Would the new water mass remain at the surface or sink? Explain.

Stability and instability of water masses

When the density of a water column increases with depth, the water column is *stable* and mixing does not occur. Conversely, when the density of a water column decreases with depth, it is *unstable*. As the dense water sinks, it produces turbulence and mixes with the layers beneath it. Instability is caused by an increase in the density of surface water due to a decrease in temperature, an increase in salinity, or both.

High evaporation rates can increase the salinity of the surface water, and low air temperatures can cool the surface water causing it to become unstable and sink. When sea ice forms near the poles, most of the salt remains in the liquid water, increasing its density and producing instability.

There is also a seasonal aspect to ocean stability. During spring and summer, stability increases as the ocean surface warms. In fall and winter, stability decreases as the ocean surface cools.

Areas of instability can produce complex patterns of stratification and thermohaline and surface circulation in the ocean.

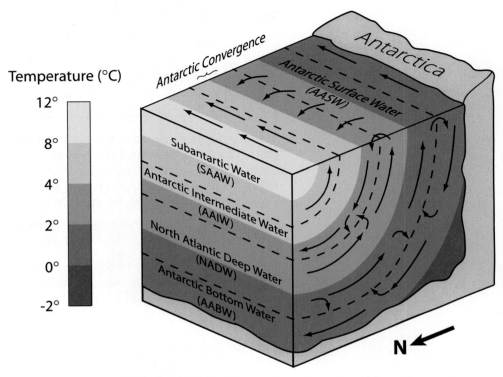

Figure 15. Thermohaline and surface currents off the coast of Antarctica. Colors represent water temperature and dashed lines represent the boundaries between water masses.

As sea ice forms along the coast of Antarctica, surface water cools and becomes more salty. This water sinks and flows northward along the ocean floor, forming the Antarctic Bottom Water mass (AABW). As winds blow the Antarctic Surface Water (AASW) eastward, the Coriolis effect deflects it toward the north. This causes upwelling of warmer, salty water, the Northern Atlantic Deep Water (NADW). This water mass mixes with the AASW to form the Antarctic Intermediate Water mass (AAIW). Because the AAIW is denser than the surface water (the Subantarctic Water mass or SAAW), it sinks below the SAAW at the Antarctic convergence.

13. Is the water column shown on the left-hand face of Figure 15 stable or unstable? Explain.

Activity 2.4

Deep-water currents

Deep-water currents, also called *density currents* or *thermohaline currents*, play a crucial role in maintaining the delicate balance of energy and nutrients in the marine environment. In this activity, you will examine two key factors that control water density—global temperatures and salinity patterns—to understand how they vary and how they affect where density currents form.

Water temperature

The first step in investigating the formation of density currents is to examine the average temperature of the ocean's surface. Temperature alters the density of sea water because water contracts when it cools. Thus, cooler water takes up less space or volume than warmer water. As a result, density increases as water temperature decreases.

Solar Flux movie

Another term for solar radiation is *solar flux*. To see an animation of how solar flux varies throughout the year, click the Media Viewer button 🎞 and choose the **Solar Radiation** movie from the media list.

Notice that the poles receive more solar radiation during their respective summers—up to 24 hours a day—than at other times of year. However, the sunlight is spread over a larger area at the poles, reducing its intensity.

▸ Launch ArcView GIS, then locate and open the **etoe_unit_2.apr** project file.

▸ Open the **Deep-water Currents** view.

The **Avg Solar Radiation (W/m^2)** theme shows the average amount of solar radiation per year, in watts per square meter, that would strike the surface if there were no clouds in the atmosphere to reflect the sunlight. Dark shades represent low amounts of radiation, and lighter shades indicate higher amounts.

▸ Use the Zoom In 🔍 and Identify tools 🛈 to examine the solar radiation data. Zoom back out to the full extent 🌐 when you are finished examining the data.

1. Describe how the average solar radiation varies with:

   a. Latitude (from pole to pole)

   b. Longitude (from east to west)

2. If there were no surface or deep-water currents circulating in the ocean, what effect might solar radiation have on the average temperature of the ocean waters?

To turn a theme on or off, click its checkbox in the Table of Contents.

To activate a theme, click on its name in the Table of Contents.

▶ Turn off the **Avg Solar Radiation (W/m^2)** theme.

▶ Turn on and activate the **Avg Surface Temperature (C)** theme.

This theme shows the average ocean surface temperature. Red represents warmer temperatures and blue represents cooler temperatures.

3. How well does the pattern of ocean surface temperature compare to the solar radiation?

4. Identify areas where surface currents may carry warm water into colder regions or cold water into warmer regions. On Map 1, place a C where ocean surface temperatures are cooler than expected and a W where they are warmer than expected. (Look for four or five of each.)

Map 1—Temperature / solar radiation anomalies and salinity extremes

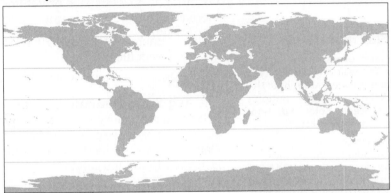

Ocean surface temperature is determined by many factors, such as solar radiation and the transfer of warm water from the equator to the poles and cold water from the poles to the equator. Deep-water convection also contributes to sea surface temperature.

▶ Turn on the **Upwelling** theme.

The **Upwelling** theme shows where deep ocean currents rise to the surface as they become less dense than the surrounding waters.

5. On Map 1, mark places where upwelling occurs with the label **U**.

▶ Compare the locations of the upwelling sites to their corresponding surface temperature anomalies (if any).

6. Are upwelling sites more likely to be associated with cold or with warm surface temperature anomalies? Explain.

Temperature anomalies

An *anomaly* is anything that is unusual, irregular, or abnormal. Ocean surface temperature anomalies occur where the surface temperature is warmer or cooler than expected for a given latitude.

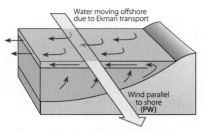

Figure 1. Conditions required for coastal upwelling.

Upwelling can occur as winds that blow parallel to the shoreline cause the surface waters to move away from the coastline, due to Ekman transport (Figure 1). Next you will examine whether winds play a role in developing these particular upwelling sites.

▶ Turn on the **Wind Direction** theme.

The **Wind Direction** theme displays the compass direction of the surface winds over both land and ocean.

▶ Examine the **Wind Direction** theme to determine the direction of the winds over the boundary currents.

7. On Map 1, mark the upwelling sites that have winds moving parallel to shore with **PW** (parallel winds).

8. Describe any patterns you observe between the upwelling sites and the coastal winds.

Upwelling sites are very important for bringing nutrient-rich water from the deep ocean to the surface where marine life can flourish.

▶ Turn off the **Upwelling** and **Wind Direction** themes.

▶ Turn on the **Density-Driven Downwelling** theme.

The **Density-Driven Downwelling** theme shows where surface water sinks as it becomes more dense than the surrounding waters. It will sink to a depth at which all the water below it is more dense and the water above is less dense.

▶ Compare the locations of density-driven downwelling sites to their corresponding surface temperatures. Use the Identify tool 🔵 to get the temperature information.

9. In what range of surface water temperatures are you most likely to find density-driven downwelling sites?

Global ocean circulation requires that if water is rising in one area, it must be sinking in another area. Next you will continue to explore the factors that influence the location of deep-water convection sites.

▶ Turn off the **Avg Surface Temperature (C)** and **Density-Driven Downwelling** themes.

Global salinity patterns

Salinity, in addition to temperature, strongly influences water density. The salts dissolved in sea water increase the density by adding to the water's mass without changing its volume. Next, you will explore global patterns of ocean salinity and the factors that influence this important property of sea water.

▶ Turn on and activate the **Avg Annual Salinity (ppt)** theme.

This theme displays the average annual salinity values for the world's oceans. Higher salinity is shown in shades of green, whereas lower salinity is shown in shades of blue.

▶ Turn on the **Major Rivers** theme.

▶ Use the Zoom In 🔍 and Identify tools 🛈 to examine the average annual salinity data. Look for patterns and anomalies in the data. Zoom back out to the full extent 🌐 when you are finished examining the data.

10. Describe how the average salinity varies with:

   a. Latitude (from pole to pole)

   b. Longitude (from east to west)

▶ Examine the regions with the highest and lowest salinity. Be sure to look in both the large oceans and the smaller seas.

11. What factors might account for the salinity levels in the areas of highest and lowest salinity? (Hint: Look at the different themes in the Table of Contents for possible factors.)

▶ Turn on the **Net Annual Evaporation (E-P, cm/yr)** theme.

This theme shows the overall loss or gain of fresh water from the ocean's surface in centimeters per year. Evaporation, in which liquid water is converted to a gas, moves fresh water from the ocean to the atmosphere. When ocean water evaporates, the salt dissolved in the water is left behind, increasing the ocean's salinity. Precipitation removes fresh water from the atmosphere and delivers it back to the ocean, reducing the ocean's salinity. Net evaporation is calculated by subtracting the total annual precipitation from the total annual evaporation.

Net Evaporation = Total Evaporation - Total Precipitation

12. Predict how you think net evaporation will affect salinity: Areas with positive net evaporation will have (circle one) higher / lower salinity than areas with negative net evaporation.

13. Where do you find areas with the:

   a. *Highest* net evaporation? Why might they be found there?

   b. *Lowest* net evaporation? Why might they be found there?

Graph 1 shows the relationship between net evaporation and ocean salinity between 60°N and 60°S. The dip near the middle corresponds to the Intertropical Convergence Zone (ITCZ), a band of moist, unstable air that circles the globe around 7°S latitude. The high net evaporation near 30°N and 30°S is due to global bands of high pressure. The cool, sinking air in these bands produces clear skies and dry conditions. Most of the world's deserts are found at these latitudes.

Graph 1—Net evaporation and surface salinity by latitude

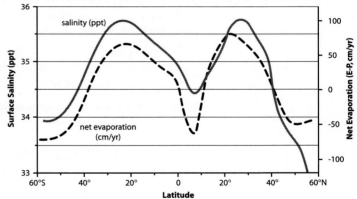

14. How well do the data in Graph 1 compare to the prediction you made in question 12?

▸ Turn off the **Net Annual Evaporation (E-P, cm/yr)** and **Major Rivers** themes.

▸ Turn on the **Density-Driven Downwelling** theme.

15. According to the legend for the **Avg Annual Salinity (ppt)** theme, what range of salinity correlates with the density-driven downwelling sites?

Seventy-five percent of the ocean has a salinity between 34 and 35 ppt. Thus, the salinity at density-driven downwelling sites is not particularly unusual. However, the temperature of seawater in these downwelling sites is definitely colder than the average sea surface temperature. How does one water mass get colder and saltier than the waters around or below it? You will explore this question by following a water molecule from the saline mid-latitude waters to the polar region of the Atlantic.

Forming deep-water masses

Evaporation creates ocean water masses with high salinity, but they are too warm to sink as density currents. In this section, you will investigate how cold, dense waters are generated.

The **Density-Driven Downwelling** theme shows several regions with significant downwelling in the North Atlantic and Southern Oceans. By measuring the surface characteristics in and around these regions, you will better understand the conditions under which density currents form.

▶ Click the QuickLoad button 🔍 and load the **North Atlantic** extent.

▶ Turn on the **Surface Currents** theme, and activate the **Avg Annual Salinity (ppt)** theme.

▶ Using the Identify tool 🛈, click on each of the points labeled B–E and record the temperature and salinity data in Table 1. (Point A has been done for you.)

Table 1—Surface water characteristics in the Northern Atlantic

Point	Temperature (°C)	Salinity (ppt)	Density (kg/m³)
A	17.5	35.6	1026.0
B			
C			
D			
E			

16. On Figure 2, plot and label the temperature and salinity values you recorded for points B through E. Use the chart to determine the density of the surface water at each point, and record the water density for each point in Table 1.

17. How does the density change from point A to point E?

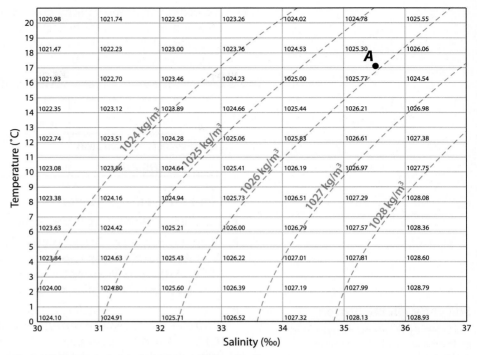

Figure 2. Temperature-Salinity-Density chart for Seawater.

How dense is fresh water?

Under normal sea-level temperature and pressure, pure fresh water has a density of 1000 kg/m³.

To create density-driven downwelling, the surface water must be denser than the water beneath it, causing the layers to overturn. This occurs in the North Atlantic, where fresh meltwater from sea ice and the Greenland ice sheet form low-density surface water masses. As warm, high-salinity surface currents flow northward into this region, they cool rapidly and become denser. When these two water masses meet in the North Atlantic, the water column becomes unstable and the higher-density, high-salinity current sinks.

The densest waters in the ocean range from 1027 to 1029 kg/m³. These correspond to a temperature of 6°C or less and a salinity of 34 ppt or higher. Next, you will perform a series of operations to identify where those waters lie and compare their density to the waters surrounding them.

▶ Click the Query button and query the **Avg Annual Salinity (ppt)** theme for the densest surface waters in the Northern Hemisphere, using these temperature and salinity values. Use the query statement:

([Hemisphere] = "N") and ([Temp (C)] <= 6) and ([Salinity (ppt)] >= 34)

▶ Set the **Found Features** option to **Highlight** and click the **New** button.

▶ Close the Query Builder window.

The densest surface waters of the Northern Hemisphere should be highlighted yellow.

▶ Click the Statistics button and calculate statistics for **the selected features of** the **Avg Annual Salinity (ppt)** theme, using the **Temp (C)** and **Salinity (ppt)** fields. (Hold down the shift key to select more than one field.) Use the **Basic** output option and click **OK**.

QuickLoad Query

If you have difficulty entering the query statement correctly:

- Click the **QuickLoad Query** button and load the **NH Dense surface waters** query.
- Select the **Highlight** option.
- Click **New**.

18. Record the average salinity and temperature inside the deep-water convection areas in Table 2. Round values to the nearest 0.1 units. Use these values to determine the average density from Figure 2 and record the density in Table 2.

Table 2—Surface water characteristics of downwelling areas

Area	Temperature (C)	Salinity (ppt)	Density (kg/m³)
Deep-water convection areas			
Adjacent waters			

▶ Close the Statistics window.

▶ Turn on the **Adjacent Waters** theme.

▶ Click the Select By Theme button  and select the features of the **Avg Annual Salinity (ppt)** theme that **Intersect all the features of** the **Adjacent Waters** theme. Click the **New** button, then close the Select By Theme dialog box.

▶ Click the Statistics button 🗵 and calculate statistics for **the selected features of** the **Avg Annual Salinity (ppt)** theme using the **Temp (C)** and **Salinity (ppt)** fields. (Hold down the shift key to select more than one field.) Select the **Basic** output option and click **OK**.

Select By Theme button

The Select By Theme button is located on the Button Bar (the top row of buttons).

19. Record the average salinity and temperature for the adjacent waters in Table 2. Use these values to determine the average density from Figure 2 and record the density in Table 2.

▶ Close the Statistics window.

20. Compare the density of water in the deep-water convection areas with that of the adjacent waters. Explain how this relates to the formation of density currents.

▶ Click the Zoom to Full Extent button 🌐 to view the entire map.

The deep-water convection area you have been studying is known as the North Atlantic Deep Water (NADW) current. Another major deep-water convection area is located in the Weddell Sea (wuh-DELL), off the coast of Antarctica, southeast of South America. This area is the source of the Antarctic Bottom Water (AABW) density current. These deep-water density currents play a major role in the distribution of nutrients throughout the world's oceans.

▶ Choose **File > Exit** to quit ArcView. Do not save changes to the project file.

Activity 2.5   # Stopping the flow

The North Atlantic Deep Water (NADW) and Antarctic Bottom Water (AABW) currents are two important parts of the *Global Conveyor Belt*. The volume of the NADW, which forms as the warm, salty Gulf Stream current moves north and cools, is believed to have a volume equal to that of 25 Amazon Rivers! The NADW cruises the deep ocean, eventually meeting up with the AABW, the densest water mass on Earth. Figure 1 shows a schematic of the entire path of the *Global Conveyor Belt* that connects surface and bottom waters in all five oceans.

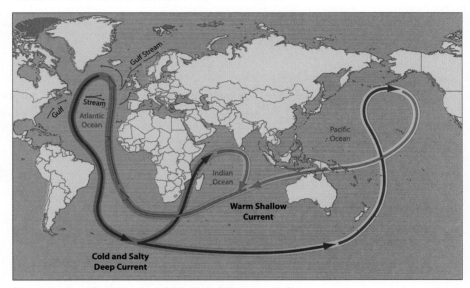

Figure 1. Simplified diagram of the Global Conveyor Belt.

1. Besides water, what else might density currents transport to the bottom of the ocean that is crucial to deep sea life?

2. The average temperature of the ocean is 3.5°C, whereas the average temperature of the ocean's *surface* is around 9°C. Does this make sense? Explain.

Scientists speculate that 12,000 years ago the *Global Conveyor Belt* shut down, and that it could do so again in the future. They hypothesize that the density currents stopped forming in response to changes in salinity.

3. Would the salinity have had to increase or decrease in order to shut down the density currents? Describe a scenario in which this could occur.

4. What types of climate changes might contribute to a shutdown of the Global Conveyer Belt? What might be some environmental consequences if the Global Conveyor Belt were to shut down?

Unit 3
Ocean-Atmosphere Interactions

In this unit, you will...

- *Explore how the distribution of land masses affects global temperature patterns*

- *Discover how the ocean and atmosphere transfer energy from equatorial regions toward the poles*

- *Compare ocean and atmospheric conditions between normal, El Niño, and La Niña phases*

- *Investigate the effects of the El Niño Southern Oscillation on climate in different parts of the world*

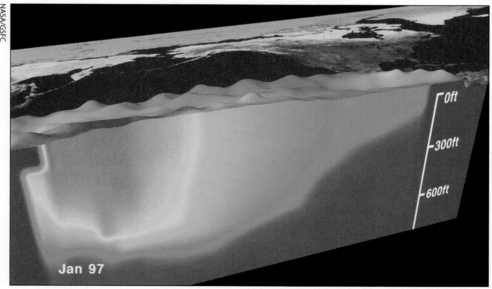

Cross-section of the equatorial Pacific Ocean in January 1997. Warm temperatures are shown in red and orange, cold in blue. Ocean depths are greatly exaggerated.

Activity 3.1

Moderating global temperature

Earth's extremes

The hottest surface air temperature ever recorded on Earth was 57.8°C (136°F), at El Azizia, Libya on September 13, 1922.

The coldest surface air temperature ever recorded on Earth was -89°C (-128.6°F), at Vostok, Antarctica on July 21, 1983.

On the moon, daytime surface temperatures at the equator average a toasty 135°C (275°F), while the poles chill at about -225°C (-375°F), a range of 360°C (650°F)! On Earth, the difference between the hottest and coldest temperatures ever recorded is only 147°C (265°F). This is still a wide range of temperatures, but not beyond what humans can endure. What is it about the oceans and atmosphere that help moderate Earth's temperature?

1. List and describe three factors about the ocean and atmosphere that you think could moderate global temperatures.

   a.

   b.

   c.

Between 40°N and 35°S, Earth receives more radiation from the sun than it gives off, and most of it is stored in the ocean. At higher latitudes, more radiation is lost than received. If the oceans and atmosphere did not circulate and redistribute this energy, sea surface temperatures would increase from the equator to the poles, forming bands of equal temperature parallel to the equator. Furthermore, average temperatures at the equator would be as much as 14°C (57°F) warmer and the poles would be up to 25°C (77°F) colder than they actually are.

Fortunately, the oceans and atmosphere do circulate, moving energy from the equator toward the poles. Next, you will examine variations in ocean temperatures and make predictions about the effect they have on land.

Energy in the balance

▸ Launch ArcView GIS, then locate and open the **etoe_unit_3.apr** project file.

▸ Open the **Sea Surface Temperatures** view.

JJA = June, July, and August

DJF = December, January, and February

The **DJF SST (C)** theme shows average sea surface temperatures during the months of December through February. Note where the warmest temperatures are located.

To turn a theme on or off, click its checkbox in the Table of Contents.

Figure 1. Compass directions.

Current directions

In general, the Coriolis effect causes currents to flow in a *clockwise* direction in the Northern Hemisphere and in a *counterclockwise* direction in the Southern Hemisphere.

▶ Turn the **JJA SST (C)** theme on and watch for changes in the locations of the warmest sea surface temperatures.

Seasonal changes in sea surface temperature may be easier to see if you view a sequence of images covering a full year.

▶ Click the Media Viewer button 🎬 and view the **SST Movie** several times. Note that the center of the map in the movie is at a different longitude than the center of the map in ArcView.

2. Describe how the location of the maximum sea surface temperature region appears to change from December–February to June–August. (In which direction does it move? See Figure 1.)

▶ Using both of the SST themes, look for areas where warmer water (red and orange) extends toward the poles. These areas represent warm currents flowing from the equator toward the poles.

3. Draw *solid* lines on Map 1 to show the locations of warm currents. Use arrowheads to indicate the direction of each current. (See **Current directions** at left.)

Map 1—Global ocean currents

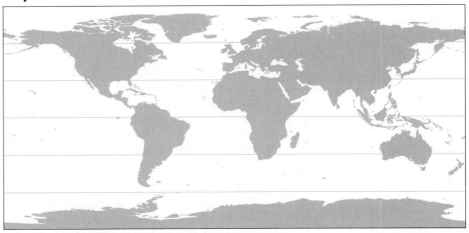

▶ Look for areas where cooler water (blue, green, and yellow) extends toward the equator. These areas represent cool currents flowing from the poles toward the equator.

4. Draw *dashed* lines on Map 1 to show the locations of cool currents. Use arrowheads to indicate the direction of each current.

Who said that?

This quote is often attributed to Mark Twain, but there is no evidence that he ever said or wrote it.

Someone once said *"The coldest winter I ever spent was a summer in San Francisco."*

5. Examine the currents near California during the summer. How might these patterns affect temperatures in San Francisco to support this quote?

6. Describe how you think life on Earth would be different if there were no ocean circulation and the equator were 14°C hotter and the poles 25°C colder than they are today.

In the next activity, you will explore how the distribution of land and ocean affects climate.

▶ Choose **File > Exit** to quit ArcView. Do not save changes to the project file.

Activity 3.2

A tale of two hemispheres

Thermal inertia, a measure of how difficult it is to change the temperature of a material, has a strong effect on how Earth stores and releases solar energy. Water has a high thermal inertia, typically four times that of soil or air. Thus, the upper layer of the ocean stores much more solar energy and it changes temperature much more slowly than does the land or the atmosphere. For this reason, the ocean is cooler during the day and warmer during the night relative to the atmosphere or land.

1. Based on your current understanding, circle the words that you think best complete the following statement:

    "The land heats up more quickly / slowly and cools off more quickly / slowly than the ocean, so the range of temperatures over the land should be greater / smaller than over the ocean."

In this investigation, you will explore temperature patterns in the atmosphere and oceans around the world to better understand the role the ocean plays in moderating global climate.

Seasonal temperature variation

▶ Launch ArcView GIS, then locate and open the **etoe_unit_3.apr** project file.

▶ Open the **Global Energy** view.

The **Seasonal Avg Air Temp Range (C)** theme shows the difference between the average surface air temperatures during June, July, and August (JJA) and the average temperatures during December, January, and February (DJF). Warm summers and cold winters result in a wide temperature range, whereas mild summer and winter temperatures or consistently cold or hot temperatures result in a narrower range.

JJA = June, July, and August

DJF = December, January, and February

Look at the distribution of the seasonal temperature ranges in the Northern and Southern Hemispheres.

2. In general, where does the seasonal temperature variation appear to be the:

    a. Largest?

    b. Smallest?

    c. Why do you think this pattern occurs?

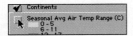

Visual estimation

To learn more about visually
estimating the percentage of
coverage, see the section on
Making visual estimations in the
Introduction to this guide.

▶ Turn off the **Seasonal Avg Air Temp Range (C)** theme.

To better understand these temperature range variations, you will
examine the distribution of land and ocean in the Northern and Southern
Hemispheres.

The distribution of land and ocean

Over 70% of the world's surface is covered by ocean, with land covering
the remaining 30%. However, the land is not distributed evenly over the
globe.

3. Visually estimate the percentage of ocean and land in each
    hemisphere and enter your estimates in Table 1.

Table 1—Estimated percentages of land and ocean by hemisphere

Hemisphere	% Land	% Ocean	Total
Northern			100%
Southern			100%

You will evaluate your estimate by calculating the actual percentage of each
hemisphere that is covered by ocean.

▶ Click the Query button 🔲 to open the Query Builder.

▶ Query the **Land and Ocean** theme for **([Hemisphere] =
"Northern") and ([Surface] = "Ocean")** as shown in steps 1–6:

QuickLoad Query

If you have difficulty entering the
query statement correctly:

• Click the QuickLoad Query
  button and load the **Northern
  Hemisphere Ocean** query.
• Select the **Highlight** option.
• Click **New**.

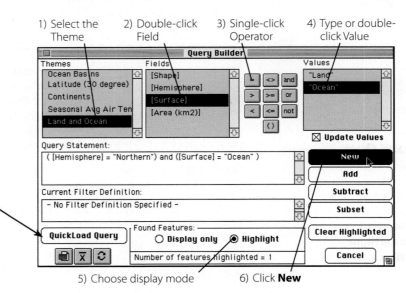

▶ Do not close the Query Builder window.

The oceans in the Northern Hemisphere should be highlighted yellow.

▶ Click the Statistics button ☒ in the Query Builder window.

▶ Calculate statistics for **the selected features of** the **Land and Ocean** theme, using the **Area (km^2)** field. Select the **Basic** output option and click **OK**.

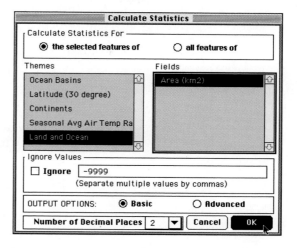

The total area of the Northern Hemisphere oceans is reported as the **Total** in the statistics window.

Rounding

To learn more about rounding, see the section on **Rounding** in the Introduction to this guide.

4. Record the total area of the oceans in the Northern Hemisphere in Table 2. Round all values to the nearest 0.1 million km².

Table 2—Ocean and land area by hemisphere

Hemisphere	Ocean Area (million km²)	Land Area (million km²)	Total Area (million km²)	% Ocean Area
Northern		100.8		
Southern		46.9		

▶ Close the Statistics window.

▶ Repeat this process to find the ocean area in the Southern Hemisphere. Record each total in Table 2, rounding to the nearest 0.1 million km².

QuickLoad Query

If you have difficulty entering the query statement correctly:

- Click the QuickLoad Query button and load the **Southern Hemisphere Ocean** query.
- Select the **Highlight** option.
- Click **New**.

([Hemisphere] = "Southern") and ([Surface] = "Ocean")

▶ Calculate the total area of each hemisphere by adding the Ocean Area and Land Area and record your results in Table 2. (Hint: The total areas of the Northern and Southern Hemispheres should be nearly equal. If not, re-check your measurements and calculations.)

Calculating % Ocean Area

$$\% \text{ Ocean Area} = \frac{\text{Ocean Area}}{\text{Total Area}} \times 100$$

▶ Calculate the % Ocean Area for each hemisphere (see left). Round your results to the nearest 0.1% and record them in Table 2.

▶ Close the Query Builder window.

Remember that land masses respond more quickly to changes in air temperature and solar radiation, so a land mass will warm or cool more quickly than the surrounding ocean.

5. Based on this information and the measurements you recorded in Table 2, in which hemisphere would you expect to find the warmest temperatures, and in which would you expect to find the coldest temperatures?

   a. Warmest

   b. Coldest

6. The daily temperature range on Earth is much smaller than that on the moon. How do you think the distribution of land and ocean affects Earth's daily temperature range?

▶ Turn off the **Land and Ocean** theme.

In the next section, you will determine which hemisphere the warmest and coldest air temperatures are found in, then examine worldwide surface air temperatures during the warmest and coldest seasons.

Air temperature

▶ Turn on and activate the **Avg JJA Air Temp (C)** theme.

The **Avg JJA Air Temp (C)** theme shows the average surface air temperatures for the months of June, July, and August (JJA)—corresponding to Northern Hemisphere summer and Southern Hemisphere winter. Reds represent higher temperatures and blues represent lower temperatures.

▶ Examine the **Avg JJA Air Temp (C)** legend (it is identical to the **Avg DJF Air Temp (C)** legend) and find areas with the highest and lowest average air temperatures at this time of year.

▶ Using the Identify tool ⬛, click on each area you found to read the average air temperature of that area.

7. Find the lowest and highest air temperatures for June through August and record the temperature and the hemisphere in which it occurs in Table 3. Round to the nearest 1°C.

Table 3—Seasonal temperature extremes

Time of year	Lowest temperature		Highest temperature		Range (°C) (highest - lowest)
	Temperature (°C)	Hemisphere (N/S)	Temperature (°C)	Hemisphere (N/S)	
June–August					
December–February					

▸ Close the Identify Results window.

▸ Turn off the **Avg JJA Air Temp (C)** theme.

▸ Turn on and activate the **Avg DJF Air Temp (C)** theme.

This theme shows average air temperatures for the months of December, January, and February (DJF)—Northern Hemisphere winter and Southern Hemisphere summer.

8. Find the lowest and highest air temperatures for December through February and record the temperatures and the hemispheres in which they occur in Table 3. Round to the nearest 1°C.

9. Calculate the temperature ranges for June–August and for December–February and record them in Table 3.

10. Using the information at left about Earth's seasons and your knowledge about thermal inertia, explain why the global temperature range is so much larger in June–August than in December–February.

Earth's seasons

Earth's seasons are:

• caused by the tilt of Earth's axis

• opposite in the northern and southern hemispheres

Dates (typical)	Hemisphere	
	Northern	Southern
Dec 21–Mar 19	Winter	Summer
Mar 19–Jun 20	Spring	Fall
Jun 21–Sep 21	Summer	Winter
Sep 22–Dec 20	Fall	Spring

▸ Turn off the **Avg DJF Air Temp (C)** theme.

Temperature range and ocean proximity

You have explored how air temperature varies by hemisphere and how the amount of ocean versus land can affect that variation. Next, you will investigate the role oceans play in moderating climate by comparing the temperature range to the distance inland from the coast.

▸ Turn on the **Temperature Range (C)** theme.

The **Temperature Range (C)** theme shows the difference between the average summer and winter daily mean temperatures for over 6000 locations around the world. Red represents large seasonal temperature differences and blue represents small seasonal temperature differences.

▶ Click the Summarize button $\boxed{\Sigma}$ to open the Summary Table dialog box.

▶ In the Summary Table Definition window, summarize the **Temperature Range (C)** theme based on the **Coastal Proximity** field. Include in the table the **Temp Range (C)** field summarized by **Average** and click the **Add** button. Choose the **All Values** option, then click **OK** to create the summary table.

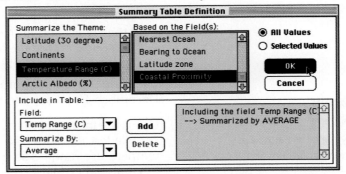

The fourth column of the summary table, labeled **Temp Range (C)_Avg,** contains the average temperature range for each of the four coastal proximity categories.

11. Record the average temperature range for each of the four proximity categories in Table 4. Round to the nearest 1°C. **(Caution: The categories in the summary table are NOT listed in the same order as in Table 4. Be sure you are entering the correct temperature range value for each category.)**

Table 4—Summary of average temperature range by coastal proximity

Coastal Proximity Category	Average temperature range (°C)
Coastal (0-80 km)	11
Near-Coastal (80–160 km)	
Mid-Coastal (160–800 km)	
Continent Interior (over 800 km)	

12. Make a bar graph of the data in Table 4.

Graph 1—Temperature range versus proximity to the coast

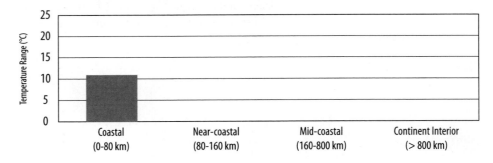

▶ Close the summary table window.

13. Based on Graph 1, describe the ocean's influence on seasonal temperature range as you move inland from the coast.

▶ Turn off the **Temperature Range (C)** theme.

Polar albedo

energy balance—the balance between incoming solar radiation and outgoing radiation.

Although the oceans are very effective at storing solar energy, the polar ice caps are excellent at reflecting solar energy. Reflectance, also called **surface albedo**, is an important component of Earth's global energy balance. Many of us have experienced the effects of albedo through the clothes that we wear. White clothes reflect solar radiation and help us stay cooler, while dark or black clothes absorb solar energy and make us feel warmer. This is true for Earth's surface as well. The north and south polar regions have different albedos, which affect their temperatures.

14. Review the temperature data for winter in each hemisphere (DJF in Northern Hemisphere and JJA in Southern Hemisphere) in Table 3. Predict which polar region you think will have the highest albedo (is most reflective), and circle the words in the statement below that summarize your prediction.

    "The coldest winter temperatures occur in the northern / southern hemisphere, which suggests that the greatest reflectance or albedo should be found in the northern / southern polar region."

▶ Turn on the **Antarctic Albedo (%)**, **Arctic Albedo (%)**, and **Permanent ice cap extent** themes.

In these themes, darker shades of brown represent lower albedos (less reflective surfaces) and lighter shades represent higher albedos (more reflective surfaces).

As you examine the Antarctic and Arctic albedo themes, note that some of the darker albedo values around both polar caps represent water during times when no sea ice is present. The permanent ice caps of Greenland and Antarctica are clearly visible, whereas the approximate extent of the permanent ice cap over the North Pole is identified by the dark blue line.

Rotate window

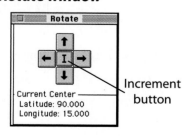

Increment button

▶ It is easier to examine the polar regions using a different map projection. To view Earth as a globe, click the Switch Projection button ▣. To center the view on the North Pole, click the Increment (I) button in the center of the Rotate window, set the **Center Point Latitude** to **90,** and click **OK**.

This view shows the albedo of the Arctic region. Try to remember what it looks like, so you can compare it to the Antarctic albedo.

▶ Click the Increment (I) button again, change the **Center Point Latitude** to **-90,** and click **OK**.

Now the view is centered on the Antarctic region.

15. Which pole appears to have the higher albedo? (Darker browns = lower albedo and lower reflectivity; lighter browns and white = higher albedo and higher reflectivity)

▶ Close the Rotate window and click the Switch Projection button 🗗 to return to a flat map view.

Now you will check your prediction by calculating the average albedo of each polar region.

▶ Click the Query button 🖾 to open the Query Builder.

▶ Select the **Arctic Albedo (%)** theme and enter the query:

([Location] = "Arctic Ocean")

▶ Set the **Found Features** option to **Highlight** and click the **New** button. Do not close the Query Builder window.

The Arctic Ocean region will be highlighted yellow.

▶ Click the Statistics button ⊠ in the Query Builder window.

▶ Calculate statistics for **the selected features of** the **Arctic Albedo (%)** theme using the **Reflectance (%)** field. Choose the **Basic** output option, then click **OK**.

The average reflectance of the Arctic Ocean region is reported in the statistics window as the **Mean**.

16. What is the average reflectance (%) of the Arctic Ocean region? Round to the nearest 1%.

▶ Repeat this process to find the average Antarctic region albedo by selecting the **Antarctic Albedo (%)** theme in the Query Builder and entering the query:

([Location] = "Antarctica")

▶ Calculate statistics for **the selected features of** the **Antarctic Albedo (%)** theme using the **Reflectance (%)** field. Choose the **Basic** output option, then click **OK**.

17. What is the average albedo percentage of the Antarctic region? Round to the nearest 1%.

▶ Close the Statistics and Query Builder windows.

18. How does the albedo you measured for each polar region compare to your prediction?

No features found?

If the **Number of features highlighted** at the bottom of the Query Builder window is 0, redo the query. Be sure to spell *Arctic* correctly!

The Southern Hemisphere is covered by land and ice in the polar region, whereas the Northern Hemisphere is covered by water and ice.

19. Does the distribution of land in the south polar region and of ocean in the north polar region increase the effect of the albedo in these regions or decrease it? Explain.

▶ Choose **File > Exit** to quit ArcView GIS, and do not save changes to the project file.

In the next activity, you will learn more about Earth's global energy balance and its impact on climate.

Activity 3.3

Climate oscillations

Our climate is the product of interactions between the atmosphere, hydrosphere (Earth's liquid water), cryosphere (glaciers and ice sheets), geosphere (solid Earth), and biosphere (living organisms) that are driven by solar energy. The ocean and the atmosphere have the greatest influence on our climate: the oceans have a strong moderating effect on global temperatures, while the atmosphere reflects harmful solar radiation and traps heat and moisture near the surface. However, all of Earth's systems are linked together, and changes to any one system can have a substantial impact on the others. To fully understand the dynamics of our climate, we must examine the global energy balance and the transfer of energy among these systems.

Global energy balance

The Sun is the primary source of energy controlling temperatures on and near Earth's surface. When solar energy reaches Earth's atmosphere, approximately 30% is reflected back into space (~25% from clouds and ~5% from the ground); 51% is transmitted to the ground or ocean surface and absorbed; and 19% is absorbed by gases in the atmosphere (Figure 1). Of the 51% of the solar energy absorbed by the surface, 7% is used to raise the temperature of the atmosphere (sensible heat flux), 21% is radiated back to the atmosphere as infrared radiation (IR), and another 23% is used to evaporate water (latent heat flux) from the ocean and soils. Ultimately, all incoming solar energy is radiated back to space—otherwise, Earth would steadily heat up.

infrared radiation (IR)—invisible form of electromagnetic radiation that is sensed as heat.

latent heat—heat energy released or absorbed during a change of phase.

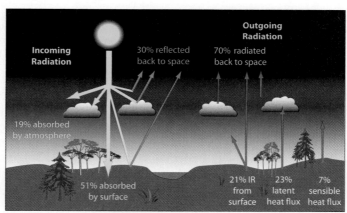

Figure 1. Annual mean global energy balance.

The total amount of solar energy reaching Earth each year is constant, but it is not distributed evenly across the globe. This uneven distribution of energy creates daily and seasonal temperature differences that drive winds, currents, precipitation, and evaporation (Figure 2).

Near the equator, the sun's rays are nearly perpendicular to Earth's surface and pass through less atmosphere, producing maximum heating.

Near the poles, more energy is absorbed by the atmosphere, and Earth's curvature causes the remaining energy to be spread over a greater area, producing less surface heating.

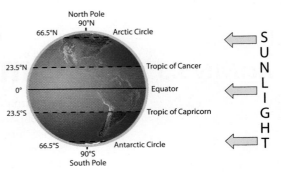

Figure 2. Variation in solar heating with latitude.

Storing energy

To understand how solar radiation affects large-scale processes such as winds, currents, and climate, you need to know a few things about water and heat energy. Water can exist in three *states*—solid, liquid, or vapor (Figure 3). When water changes states, it absorbs or releases more heat (called *latent heat*) than do most other substances.

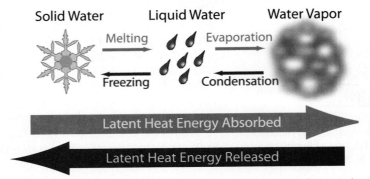

Figure 3. As ice melts or water evaporates, it absorbs latent heat, and as water vapor condenses or water freezes, it releases latent heat. The transition between liquid water and water vapor involves about seven times as much heat energy as the transition between liquid water and ice.

heat capacity—the amount of energy required to raise the temperature of an object by 1°C.

Water can absorb more energy than other substances because it has a high *heat capacity*. Raising the temperature of water requires a lot of energy, so it is very difficult to change the temperature of the ocean even a small amount. This resistance to change is called *thermal inertia*. However, even small changes to the energy content of the ocean can have considerable effects on global climate. In contrast, the heat capacity and thermal inertia of the atmosphere is much lower, making it easier to change the temperature of the atmosphere.

At the equator, where solar energy is highest, about 75% of Earth's surface is covered by water. Water's higher heat capacity allows the ocean to absorb and retain more solar energy than the land or atmosphere. In fact, the upper few hundred meters of ocean stores approximately 30 times more heat than the entire atmosphere. Without some type of circulation in the ocean and atmosphere, the equator would be 14°C (57°F) warmer on average than now, and the North Pole would be 25°C (77°F) colder. Fortunately, temperature imbalances drive atmospheric and oceanic circulation, which redistribute energy more evenly over Earth's surface.

1. Today, 75% of the area at the equator is covered with water. How might global temperatures change if 75% of the area near the equator were covered with land? Explain.

Next, you will examine other ways that Earth keeps its cool.

Reflecting energy

Earth's ability to store solar energy is critical for regulating global temperatures. Equally important is its ability to reflect solar energy. The reflectivity of a material—the ratio of the amount of light reflected by an object to the amount of light that falls on it—is called the *albedo*. The albedo is determined principally by a material's texture and color; rougher, darker surfaces tend to have lower albedos, whereas smooth, bright surfaces reflect more efficiently and have higher albedos. In addition, the angle at which the sun's rays strike the surface affects the albedo.

Albedo of surface types at 45°N latitude

Surface Type	Albedo (%)
dense swampland	9–14%
deciduous trees	13%
grassy field	20%
barren field	5–40%*
farmland	15%
desert or large beach	25%*

* Depending on color of surface.

The albedos of various surface materials range widely, from extremely reflective to highly absorptive. For instance, different states of water—as ice, liquid water, and water droplets—have very different albedos. Ice and certain types of clouds have a high albedo and reflect 60–90% of the sun's energy. The oceans, on the other hand, generally have a low albedo (10%), which means that they absorb 90% of the sun's energy that strikes them. The albedo of the biosphere, meanwhile, varies according to the type of land cover; vegetated land has a much lower albedo than barren land. Thus, tropical rainforests absorb more solar energy than deserts.

The average albedo for the whole Earth is about 30%. This is not a permanent characteristic; any significant alteration of Earth's surface, such as melting of the ice caps, desertification, or burning of tropical rain forests, can trigger changes in the amount of solar radiation absorbed or reflected. A change in albedo is significant, because global albedo is a key determinant of conditions on Earth. For example, if the ice sheets melted, the global albedo would decrease. Earth's surface would then absorb more energy, and the temperature of the atmosphere would rise. Evidence in the geologic record suggests this happened during the Cretaceous Period (144 to 65 million years ago). At that time, there was little or no snow and ice cover, even at the poles, and global temperatures were at least 8–10°C warmer than today. Such global warming hinders the formation of sea ice and disrupts the development of cold, dense currents that sink to the depths of the ocean. In addition, warmer temperatures melt the ice sheets. This, in turn, increases the volume of the oceans, raises global sea levels, and triggers changes in Earth's energy balance.

Circulating energy with winds and currents

The uneven distribution of solar energy produces circulation in the ocean and atmosphere that moderates temperatures across the globe. Near the equator, evaporation adds large amounts of water vapor and latent heat to the atmosphere (Figure 4). Hot, moist air rises from the ocean surface and cools, causing some of the water vapor to condense and fall back to Earth as rain. As water vapor condenses, it releases latent heat. This generates temperature differences and winds higher in the atmosphere, which transport the remaining water vapor and latent heat toward the poles. As the air moves toward the poles, it continues to cool, becomes denser, and descends, flowing back towards the equator to complete the cycle.

Ocean currents contribute to the heat transfer between the equator and the poles by moving warm water toward the poles and cold water toward the equator. Warm surface currents at the equator are driven westward by the winds and deflect toward the poles as they near continents. As these warm currents flow toward the poles, they transfer heat to the atmosphere, warming the atmosphere and cooling the currents. The cooled surface currents then flow back toward the equator to be warmed again.

Deep-water density currents also play an important role in redistributing energy. Near the poles, cold, salty waters sink to great depths and flow slowly toward the opposite pole. These currents eventually rise to the surface where the water is reheated to begin the cycle again.

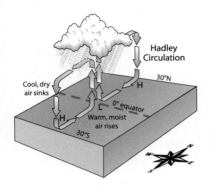

Figure 4. Simple convection cells between the equator and 30°N and 30°S move energy from the equator toward the poles. This process is called *Hadley circulation*.

El Niño and the Southern Oscillation

The ocean's high thermal inertia (slow response to change) helps stabilize Earth's climate. However, this stability is not a constant. Over the past century, scientists have observed *oscillations* in sea surface temperature and air pressure that occur over years and decades. Localized ocean temperature increases of only 2–3°C have been linked to extended periods of drought and flooding globally. Heat transfer between the ocean and atmosphere, which causes changes in surface air pressure and winds, drives the oscillations. The best known of these is the *Southern Oscillation*, a seesaw shift in surface air pressure driven by changes in sea surface temperature between the eastern and western Pacific Ocean in the Southern Hemisphere.

oscillation—back and forth pattern of change between one state or direction and its opposite.

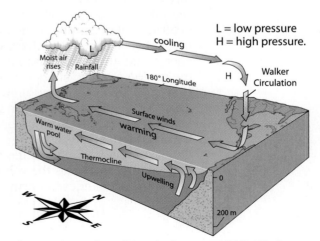

Figure 5. Normal conditions in the equatorial Pacific Ocean.

What drives the Southern Oscillation?

temperature gradient—change in temperature over a distance.

Strong evaporation and convection at the ocean's surface drive the *Southern Oscillation*. In a normal year, the average sea surface temperature is warmer in the western Pacific than in the eastern Pacific (Figure 5). This *temperature gradient* causes air to circulate parallel to the equator in a pattern called the *Walker Circulation*. Warm waters of the western Pacific produce high levels of evaporation and increased precipitation. The warm, rising air creates low pressure at the surface. As surface winds blow into this low pressure center, they pile up water to form a small mound in the western Pacific. As a result, the layer of warm surface water is deeper in the western Pacific than in the east. This forces the deeper and colder waters in the western Pacific to flow eastward and up toward the surface, producing cooler sea surface temperatures off the coast of South America. The rising, moist air in the western Pacific fuels the heavy monsoon rains of Southeast Asia, and the sinking dry air in the eastern Pacific creates the coastal deserts of South America.

2. Circle the word in each pair that correctly completes the sentence: "Warmer / Cooler surface temperatures cause increased evaporation and upward convection of air masses, which leads to higher / lower air pressure near the surface."

Changes to the Southern Oscillation—El Niño and La Niña

phase—stage in a periodic process or phenomenon.

About every three to eight years, the Southern Oscillation strengthens or weakens and we experience either El Niño or its sister phase, La Niña. During an El Niño phase (Figure 6), a 2–3°C warming of the water in the eastern Pacific reduces the sea surface temperature gradient from east to west. The lower temperature gradient weakens the east to west trade winds in the south Pacific. This allows the mound of warm water piled up in the western Pacific to spread out and move eastward. The warm surface waters in the east and high air pressures prevent the typical upwelling of cold, nutrient-rich water off the west coast of South America.

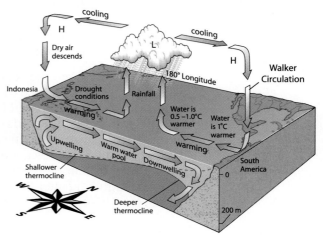

Figure 6. Conditions in the equatorial Pacific Ocean during the El Niño phase.

Although the El Niño Southern Oscillation occurs in the tropical Pacific Ocean, its impact on climate is observed globally. Variations in rainfall are significant, with reports of droughts in regions of Indonesia and Australia that are normally drenched by monsoons, and storms and flooding in Ecuador and parts of the United States. In addition to climate changes, El Niño phases can have devastating effects on fisheries, agricultural productivity, and outbreaks of fire and disease. For example, in most years colder nutrient-rich water from the deeper ocean is drawn to the surface near the coast of Peru and Ecuador (upwelling). This produces abundant plankton, the food source of the anchovy, and the anchovy harvest increases. However, when upwelling weakens in an El Niño phase and warmer low-nutrient water spreads along the coast, the anchovy harvest plummets.

The El Niño phase that produces warming of the water in the eastern Pacific basin is counteracted in other years by a similar cooling phase known as La Niña. During the La Niña phase, trade winds are anomalously stronger than normal, and the sea surface temperature gradient increases (Figure 7). This causes greater upwelling of cold water in the eastern Pacific Ocean and increased warming and evaporation in the western Pacific Ocean.

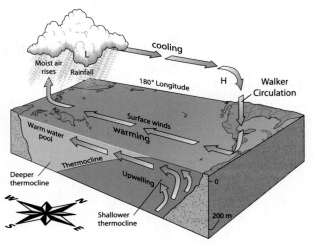

Figure 7. Conditions in the equatorial Pacific Ocean during a La Niña phase.

These oscillations and the conditions that produce them are the focus of extensive research because their climate impacts can be quite severe. A recent study suggests that El Niño can be triggered by volcanic eruptions. Climate and eruption records dating back to the seventeenth century suggest that a large eruption can double the likelihood of an El Niño phase the following winter. Volcanoes might alter the climate by spewing great quantities of dust and greenhouse gases into the air. The dust particles reflect sunlight and alter the amount of solar heat reaching Earth's surface.

3. List three ways in which El Niño might affect your everyday life. Consider both direct and indirect effects.

   a.

   b.

   c.

Other climate oscillations

The El Niño Southern Oscillation is the most studied and best understood ocean-atmosphere interaction. In recent decades, other patterns have been identified around the globe, including the Pacific Decadal Oscillation and the Atlantic Multidecadal Oscillation. The Pacific Decadal Oscillation is the primary factor determining variations in monthly sea surface temperature in the North Pacific (poleward of 20°N). The Pacific Decadal Oscillation has a distinct pattern of long-term sea surface temperature oscillations in the North Pacific. However, it occurs over 20–30 years, whereas typical El Niño Southern Oscillation phases last for only 6–18 months.

A similar pattern of sea surface temperature oscillations in the North Atlantic Ocean—the Atlantic Multidecadal Oscillation—occurs over a period of 50-80 years. Scientists believe that the Atlantic Multidecadal Oscillation and Pacific Decadal Oscillation are linked to the major droughts of the past 100 years in North America including the Dust Bowl in the 1930s, extended droughts in the 1950s and 1970s, and the current drought in the southwestern United States that started in 1998. Understanding these ocean-atmospheric phenomena is critical to understanding global climate variability and improving climate-forecasting capabilities.

Activity 3.4

El Niño and La Niña

For at least the past 400 years, people living around the equatorial Pacific Ocean have noted that in some years the surface waters of the central and eastern Pacific Ocean are warmer than usual. Because this change usually begins in December, the phenomenon was named El Niño—Spanish for "the Christ child." Over time, scientists have come to realize that El Niño is not an isolated phenomenon. Subsequent years sometimes show a pattern of unusually cool surface waters in this region, an event now referred to as the La Niña phase. Both phases can have dramatic effects on global climate. Too wet, too dry, too hot, or too cold—during an El Niño or La Niña phase, chances are good that the weather will be unusual.

phases—stages in a periodic process or phenomenon.

Normal phase

In the first part of this investigation, you will examine typical ocean and atmospheric circulation patterns in the Pacific Ocean under normal phase conditions. Later, you will compare these patterns to El Niño and La Niña phases to see how small changes can disrupt local circulation patterns and global weather conditions.

▶ Launch ArcView GIS, then locate and open the **etoe_unit_3.apr** project file.

▶ Open the **Normal Phase** view.

The **Wind Speed (m/s)** shows the average speeds and directions of surface winds around the globe. Arrows point in the average wind direction (Figure 1), and the color of each arrow represents the average wind speed at that location. Notice how, in this theme as well as many of the others, most characteristics of the ocean and atmosphere change more with latitude than with longitude.

▶ Click the QuickLoad button 🔲 and load the **Pacific Ocean** extent.

1. Examine the surface winds over the Pacific Ocean between the equator and 30° in both hemispheres and describe any differences you observe:

    a. *between* the Northern and Southern Hemisphere.

    b. *within* each hemisphere, from east to west.

Figure 1. Compass directions.

The winds blowing from east to west near the equator are called the Trade Winds, named for their importance in pushing the sailing ships that traveled between Europe and the Americas in the 17th and 18th centuries.

To turn a theme on or off, click its checkbox in the Table of Contents.

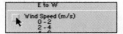

▶ Turn off the **Wind Speed (m/s)** theme and turn on the **Surface Currents** theme.

The **Surface Currents** theme shows the average directions of the world's ocean surface currents. Again, note the directions of the ocean currents between 30°N and 30°S.

2. Between 30°N and 30°S, in what direction do most surface currents flow? How does this compare to the average direction of the Trade Winds?

Next, you will look at the average sea surface temperatures (SST) across the Pacific Ocean.

▶ Turn off the **Surface Currents** theme.

▶ Turn on the **Average SST (C)** theme.

Examine the sea surface temperatures across the Pacific Ocean, paying particular attention to the region between 30°N and 30°S.

3. Which part of the equatorial Pacific Ocean appears to have the warmer surface temperature, the eastern Pacific or the western Pacific?

Water expands as it warms and contracts as it cools. This phenomenon is called *thermal expansion*.

4. Based on the average surface temperature alone, what effect do you think this warmer water would have on sea level in this part of the Pacific?

▶ Turn off the **Average SST (C)** theme and turn on the **Sea Level Anomaly (cm)** theme.

Sea not-so-level

You may have been taught that sea level is a flat surface that extends all over the ocean, but in fact the ocean surface has many small hills and valleys that vary from mean sea level by tens to hundreds of centimeters. These are caused by temperature and pressure differences, tides, gravity, and the effects of wind blowing across the ocean surface.

An *anomaly* is anything that deviates from normal or average. In this theme, areas where sea level is higher than expected are colored pink, and areas that are lower than expected are colored blue. White represents places where sea level is at its expected normal height of 0 meters. Next, you will see if the sea level anomaly data support your prediction.

▶ Turn on the **Eastern Pacific** and **Western Pacific** themes.

These themes define the regions you will use to compare the eastern and western equatorial Pacific Ocean. You will begin by using the **Sea Level Anomaly (cm)** theme to look for differences in sea level between the two areas.

Select By Theme button

The Select By Theme button is located on the Button Bar (the top row of buttons).

▶ Click the Select By Theme button 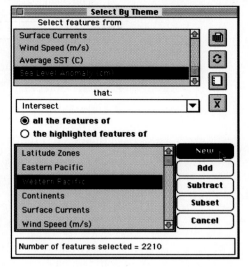 (location shown at left) and select features from the **Sea Level Anomaly (cm)** theme that **Intersect all the features of** the **Western Pacific** theme.

▶ Click the **New** button but do not close the Select By Theme window. The **number of features selected** is reported at the bottom of the dialog box. It should read **2210**.

▶ Click the Statistics button $\boxed{\overline{x}}$ in the Select By Theme window and calculate statistics for **the selected features of** the **Sea Level Anomaly (cm)** theme using the **Anomaly (cm)** field. Select the **Basic** output option and click **OK**.

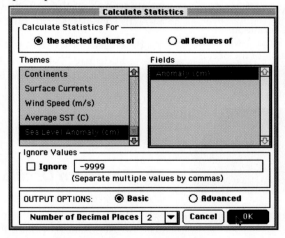

The average sea level anomaly in centimeters is given as the **Mean** in the statistics window.

5. Round the average sea level anomaly for the region to the nearest 0.1 cm and record it in Table 1.

Table 1—Sea level anomalies in the equatorial Pacific Ocean

Western Pacific Anomaly (cm)	Eastern Pacific Anomaly (cm)	Difference (W Pacific Anomaly – E Pacific Anomaly) (cm)

▶ Close the Statistics window.

▶ In the Select By Theme window, click the Clear Selection button ▣.

▶ Repeat these procedures to determine the average sea level anomaly for the eastern Pacific and record the value in Table 1. (The number of selected features should be **2179**.)

6. Calculate the difference between the average sea surface anomaly in the western Pacific and the eastern Pacific and record it in Table 1.

▶ When you have completed Table 1, close any Statistics or Select By Theme windows that are still open.

7. According to Table 1, is sea level higher in the western Pacific or in the eastern Pacific? Compare this result with your prediction in question 4.

Pacific Ocean profile

To understand the conditions in the Pacific Ocean, it may help to look at a cross section of the ocean and surface winds. Click the Media Viewer button 🎞 and open the **Pacific Ocean Profile** image.

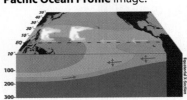

This difference represents a low but very broad "hill" of water in the western half of the Pacific Ocean. This hill is essentially permanent, even though gravity is constantly pulling the hill down and outward.

8. Explain how the prevailing winds and ocean currents might help keep the "hill" of water in the Pacific Ocean from spreading out. (See **Pacific Ocean profile** at left.)

The Southern Oscillation

Long before they understood the causes of El Niño, scientists noticed seemingly independent weather phenomena in the Southern Hemisphere. One of these was an oscillation, or periodic strengthening and weakening, of the surface atmospheric pressure across the Pacific Ocean. Scientists now understand that this oscillation and the sea surface temperature changes characteristic of El Niño and La Niña phases are directly related. Together, these phenomena are now known as the *El Niño Southern Oscillation*, or ENSO.

In this section, you will compare the average atmospheric pressure in the equatorial eastern and western Pacific Ocean for three different oscillation phases: the "normal" long-term average calculated with data from *all* years between 1950 and 2003, eight years of El Niño, and eight years of La Niña.

millibar (mb)—a bar (from the Greek *báros*, meaning *weight*) is a unit of pressure, equal to ten newtons per square centimeter. *Milli* is the standard metric prefix for *one-thousandth*, so one millibar equals 0.001 bar. Standard atmospheric pressure is 1013.2 mb.

▶ Click the QuickLoad button 🔍 and load the **Southern Oscillation** view.

The **Normal Surface Pressure (mb)** theme represents the average surface pressure for all years from 1950 to 2003. Low pressure is represented by light shades of green, and high pressure is shown in darker shades of green. Surface pressure decreases as the air temperature and humidity increase.

▶ Click the QuickLoad button 🔍 and load the **Pacific Ocean** extent.

▶ Click the Select By Theme button 🔲 and select features from the **Normal Surface Pressure (mb)** theme that **Intersect all the features of** the **Eastern Pacific** theme.

▶ Click the **New** button but do not close the Select By Theme window.

▶ Click the Statistics button $\boxed{\overline{x}}$ in the Select By Theme window and calculate statistics for **the selected features of** the **Normal Surface Pressure (mb)** theme using the **Pressure (mb)** field. Select the **Basic** output option and click **OK**.

9. Round the average (**Mean**) surface pressure for the normal phase to the nearest 0.1 mb and record it in Table 2.

Table 2—Equatorial Pacific average surface pressure and climate phases

Region	Average surface pressure (mb)		
	Normal phase	El Niño phase	La Niña phase
Western Pacific	1009.7	1010.1	1009.2
Eastern Pacific			
Difference (E-W)			

▶ Close the Statistics window.

▶ Repeat these steps using the **El Niño Surface Pressure (mb)** and **La Niña Surface Pressure (mb)** themes and record the average surface pressure for the eastern Pacific in Table 2.

▶ Close the Select By Theme window.

10. Calculate the differences between the eastern and western Pacific regions for the three phases and record them in Table 2.

Winds blow from areas of high pressure to areas of low pressure, and the wind speed increases as the pressure differences increase. With this in mind, consider how the pressure difference between the western and eastern Pacific Ocean influences wind speeds during El Niño and La Niña phases.

11. Write the name *El Niño* or *La Niña* in each blank to correctly complete the following statement:

"The pressure difference between the eastern Pacific and the western Pacific is greatest during the _____ phase, resulting in higher wind speeds than in the normal or _____ phase.

▶ Turn off the **Normal Surface Pressure (mb)** theme.

Next, you will measure wind speeds in the equatorial Pacific Ocean to see whether wind speeds during the El Niño and La Niña phases respond to the east-west pressure difference in the way you predicted.

▶ Turn on the **Wind Speed (m/s)** theme.

▶ Click the Select By Theme button 🔲 and select features from the **Wind Speed (m/s)** theme that **Intersect all the features of** the **Eastern Pacific** theme.

▶ Click the **New** button, but do not close the Select By Theme window.

This process highlights the average wind speed data for the eastern Pacific. Next, you will add the data for the western Pacific so you can calculate the average wind speed across the entire equatorial Pacific.

▶ In the Select By Theme window, select features from the **Wind Speed (m/s)** theme that **Intersect all the features of** the **Western Pacific** theme, then click the **Add** button. (NOT the **New** button!) Do not close the Select By Theme window.

Both the eastern and western equatorial Pacific Ocean should be highlighted yellow, and the **number of features selected** should read **656**.

▶ Click the Statistics button ⊠ in the Select By Theme window and calculate statistics for **the selected features of** the **Wind Speed (m/s)** theme, using the **Normal Mean Wind Speed, El Niño Mean Wind Speed**, and **La Niña Mean Wind Speed** fields. (Hold down the Shift key to select multiple fields.) Select the **Basic** output option and click **OK**.

12. Round the average (**Mean**) wind speeds to the nearest 0.01 m/s and record them in Table 3.

Table 3—Equatorial Pacific Ocean wind speed comparison

Average wind speed (m/s)		
Normal phase	*El Niño phase*	*La Niña phase*

▶ Close the Statistics and Select By Theme windows.

13. Do the average wind speeds you recorded in Table 3 agree with the predictions you made in question 11 about the wind speeds in La Niña years? Explain.

The effects of El Niño and La Niña on sea level

Now that you have seen how surface atmospheric pressure and wind speed vary during El Niño and La Niña phases, think back to the concept of sea level you explored earlier. As you observed, the warm waters of the equatorial western Pacific form a hill or mound due to thermal expansion, and are "held" in place by the trade winds and equatorial currents. You have also seen that the trade winds strengthen or weaken in response to changes in surface pressure and ocean temperature associated with El Niño and La Niña phases.

Next you will examine how the strengthening or weakening of the trade winds affects sea level.

▶ Turn off the **Wind Speed (m/s)** theme.

▶ Turn on the **Sea Level Stations** theme.

This theme shows the locations of sea level monitoring stations in the Pacific Ocean. The points in the middle of the Pacific are located on islands.

▶ Click the QuickLoad button 🔲 and load the **El Niño Sea Level Anomalies** legend.

The sea level stations are now classified based on whether sea level during El Niño phases is below normal (blue) or above normal (red). In this case, normal is the mean sea level at that location.

14. Based on what you have learned about the effect of El Niño on the trade winds and sea level, complete the following statement by circling the appropriate words where indicated and writing *higher* or *lower* in each blank.

    During an El Niño phase, the speed of the Trade Winds is _____ than in the normal phase, causing the "hill" of warm water in the western Pacific to pile up / spread out (circle one). As a result, sea level is _____ than normal in the western Pacific and _____ than normal in the eastern Pacific.

▶ Click the QuickLoad button 🔲 and load the **La Niña Sea Level Anomalies** legend.

15. Based on what you have learned about the effect of La Niña on the trade winds and sea level, complete the following statement by circling the appropriate words where indicated and writing *higher* or *lower* in each blank.

    During La Niña phases, the speed of the Trade Winds is _____ than in the normal phase, causing the "hill" of warm water in the western Pacific to pile up / spread out (circle one). As a result, sea level is _____ than normal in the western Pacific and _____ than normal in the eastern Pacific.

▶ Turn off the **Sea Level Stations** theme.

Effects of El Niño and La Niña on sea surface temperature

Now you will explore how El Niño and La Niña affect ocean surface temperatures.

▶ Turn on the **El Niño SST Anomaly (C)** theme.

This theme shows sea surface temperature anomalies averaged over eight years of the El Niño phase. Reds represent positive anomalies, where the temperature was higher than normal, and blues represent negative anomalies, where the temperature was lower than normal.

16. In the **El Niño SST Anomaly (C)** theme, where are large positive sea surface temperature anomalies (warming) more common—in the eastern or the western Pacific Ocean?

▶ Turn off the **El Niño SST Anomaly (C)** theme and turn on the **La Niña SST Anomaly (C)** theme.

17. In the **La Niña SST Anomaly (C)** theme, where are large negative temperature anomalies (cooling) more common—in the eastern or in the western Pacific Ocean?

Now you will quantify the sea surface temperature anomalies across the equatorial Pacific Ocean.

▶ Click the Select By Theme button 🔲 and select features from the **El Niño SST Anomaly (C)** theme that **Intersect all the features of** the **Western Pacific** theme.

▶ Click the **New** button but do not close the Select By Theme window.

▶ Click the Statistics button ☒ in the Select By Theme window and calculate statistics for **the selected features of** the El Niño SST Anomaly (C) theme using the **SST Anomaly (C)** field. Select the **Basic** output option and click **OK**.

18. Record the average (**Mean**) SST anomaly for the region in Table 4. Round to the nearest 0.01°C.

Table 4—SST anomalies for El Niño and La Niña phases

Phase	SST Anomaly (°C)	
	Western Pacific	Eastern Pacific
El Niño		
La Niña		

▶ Repeat these procedures to determine the SST anomaly for the **El Niño SST Anomaly (C)** theme in the **Eastern Pacific** and the **La Niña SST Anomaly (C)** theme in the **Western Pacific** and the **Eastern Pacific** to complete Table 4.

▶ When you are finished, close the Statistics and Select By Theme windows.

19. Warming of the ocean surface produces increased evaporation and precipitation, whereas cooling decreases evaporation and precipitation. Based on the data recorded in Table 4:

a. Which part of the Pacific Ocean should experience the largest increase in precipitation and evaporation? During which phase of the oscillation does this occur? (Normal, El Niño, or La Niña)

b. Which part of the Pacific Ocean should experience the largest decrease in precipitation and evaporation? During which phase of the oscillation does this occur? (Normal, El Niño, or La Niña)

▶ Turn off the **La Niña SST Anomaly (C)** theme.

Weather and climate effects

In addition to producing predictable weather and climate patterns in the equatorial Pacific, the El Niño Southern Oscillation also influences weather and climate around the world. To illustrate this, you will examine the effects of El Niño and La Niña on global precipitation rates and drought patterns.

▶ Click the Zoom to Full Extent button 🌐 to view the entire map.

PPT—precipitation.

▶ Turn on the **El Niño PPT Anomaly (cm/yr)** theme.

This theme shows the deviation from normal precipitation averages during the El Niño phase, in centimeters per year. Negative anomalies (below average values) are shown in orange, and positive anomalies (above average values) are shown in purple.

▶ Turn off the **El Niño PPT Anomaly (cm/yr)** theme and turn on the **La Niña PPT Anomaly (cm/yr)** theme.

This theme shows the deviation from normal precipitation averages during La Niña phase, in centimeters per year.

20. Based on the maps of precipitation anomalies, what can you say about global precipitation patterns during the El Niño and La Niña phases?

▶ Turn off the **La Niña PPT Anomaly (cm/yr)** themes.

ENSO and climate

To conclude your investigation, you will examine two themes that indicate whether the worldwide climate conditions were unusually dry or wet during El Niño or La Niña phases.

▶ Turn off the **Continents** theme.

▶ Turn on the **El Niño PDSI** and the **Countries** themes.

PDSI units?

PDSI is a numerical value calculated using several climate measurements, but has no units itself. It is useful for comparing conditions of different regions or of the same region at different times.

To activate a theme, click on its name in the Table of Contents.

The Palmer Drought Severity Index (PDSI) is a measure of the relative wetness or dryness of an area. This theme shows the PDSI values for areas around the world. This is also an anomaly theme—negative values, in brown, represent drought conditions; and positive values, in green, represent wetter than normal conditions.

21. Identify three countries that appear to be wetter than average during El Niño phases, and three that appear to be drier than average. (If you don't know the name of a country, activate the **Countries** theme and click on the country using the Identify tool 🛈.

    a. Wetter than average countries.

       1)

       2)

       3)

    b. Drier than average countries.

       1)

       2)

       3)

Next, you will find the average PDSI values for six selected countries to assess how each is affected during an El Niño phase.

▶ Click the QuickLoad button 🔍 and select the **El Niño PDSI by Country** query. This method will skip the Query Builder window to automatically run the following query:

([Name] = "United States") or ([Name] = "Argentina") or ([Name] = "Australia") or ([Name] = "Finland") or ([Name] = "South Africa") or ([Name] = "India")

The data collection sites in the six selected countries should be highlighted yellow. Now you will average the data for each country by creating a summary table.

▸ Click the Summarize button **Σ**.

▸ In the Summary Table Definition window, summarize the **El Niño PDSI** theme based on the **Name** field. Include in the table the **PDSI** field summarized by **Average** and click the **Add** button. Choose the **Selected Values** option, then click **OK**.

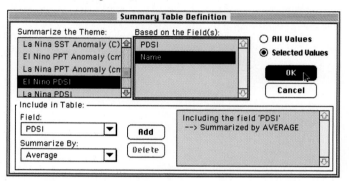

22. Round the average El Niño PDSI values for the six countries to the nearest 0.01 and record them in Table 5.

Table 5—El Niño and La Niña PDSI values for selected countries

Country Name	PDSI Value		
	El Niño	La Niña	Change +/-
Argentina			
Australia			
Finland			
India			
South Africa			
United States			

▸ Close the summary table window.

▸ Turn off the **El Niño PDSI** theme and turn on the **La Niña PDSI** theme.

▸ Click the QuickLoad button 🔍 and select the **La Niña PDSI by Country** query.

▸ Click the Summarize button **Σ** and summarize the **La Niña PDSI** theme based on the **Name** field. Include in the table the **PDSI** field summarized by **Average** and click the **Add** button. Choose the **Selected Values** option, then click **OK**.

23. Round the La Niña PDSI results to the nearest 0.01 and record them in Table 5.

24. Based on your results, which of the six countries are most likely to experience:

   a. drought during El Niño years?

   b. wet La Niña years ?

25. Summarize how the general patterns in drought and wet years vary with El Niño and La Niña.

▶ Choose **File > Exit** to quit ArcView GIS, and do not save changes to the project file.

Activity 3.5

Local interactions

Now that you have examined some of the global effects during El Niño and La Niña phases, you will look at how these events affect the temperature and precipitation across the United States in general, and at a specific location of your choice.

▶ Launch ArcView GIS, then locate and open the **etoe_unit_3.apr** project file.

▶ Open the **Regional Effects** view.

The **US Climate** theme contains average precipitation and temperature data for different climate regions of the United States during normal, El Niño, and La Niña phases. In order to visualize these data, you will load a series of legends, each showing a different climate measurement.

Loading saved legends

To load a saved legend, click the QuickLoad button and load the specified legend file.

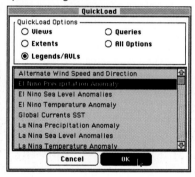

1. Click the QuickLoad button and load the following legends for the **US Climate** theme, one at a time. For each legend, describe the pattern it shows across the United States. For temperature, describe states or regions that are much warmer or cooler than average; for precipitation, describe areas that are much wetter or drier than average.

    a. El Niño Precipitation Anomaly

    b. El Niño Temperature Anomaly

    c. La Niña Precipitation Anomaly

    d. La Niña Temperature Anomaly

▶ Choose a location in the United States and use the Zoom In tool to zoom in on it.

▶ With the **US Climate** theme active, click on your chosen location using the Identify tool to view climate information about that location.

2. Write the name of your chosen location in Table 1, then use the information in the Identify Results window to complete the table. Round the temperature and precipitation anomaly values to the nearest 0.01 unit.

Table 1—Climate data for your chosen location

Information	Value
Name	
State	
(Normal) Ave Annual Precip (cm/yr)	
(Normal) Ave Annual Temp (°C)	
El Niño Precip Anomaly (cm)	
El Niño Temp Anomaly (°C)	
La Niña Precip Anomaly (cm)	
La Niña Temp Anomaly (°C)	

3. Using the average El Niño, La Niña, and Normal values from Table 1, describe how each of the following phases affects the climate of your chosen location. (Temperature, precipitation, etc.)

   a. El Niño

   b. La Niña

Present conditions

Broken links?

If the web links in this activity do not work, alternate links may be found at:

http://www.scieds.com/saguaro/ etoe/act35.html

To open the web page when the **etoe_unit_3.apr** project file is open:

- Click the Media Viewer button.
- Load the **NOAA El Nino** web page.

NOAA El Niño Page

http://www.elnino.noaa.gov

Now that you have learned about the effects that El Niño and La Niña have on your chosen area, you will see what the present conditions are in the South Pacific Ocean.

▶ Open the NOAA El Niño web page using the QuickLoad button or by entering its address directly in your web browser:

http://www.elnino.noaa.gov

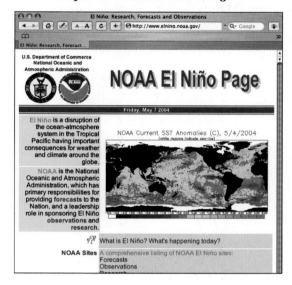

The NOAA El Niño web page displays a map of the present sea surface temperature anomalies. Examine the temperature anomalies as they exist today, and think about what you have learned about El Niño and La Niña phases.

4. Describe the pattern of temperature anomalies. Is it unusually warm in the east or west part of the equatorial Pacific Ocean, or are the temperatures as expected for normal phase years?

El Niño Forecasts

http://www.elnino.noaa.gov/ forecast.html

Current El Niño status and US forecasts

http://www.cpc.ncep.noaa.gov/ products/analysis_monitoring/ lanina/

▶ Click the **Forecasts** link, then click the **Current El Niño status and US forecasts** link on the El Niño Forecasts page.

This web site provides additional information on present sea surface temperatures and temperature anomalies. Use this information and the information on the main El Niño web page to answer the following question.

5. Based on the sea surface temperature anomalies as shown on the NOAA web site, do present conditions resemble an El Niño phase, a La Niña phase, or normal phase conditions?

Current Diagnostic Discussion

http://www.cpc.ncep.noaa.gov/products/analysis_monitoring/enso_advisory/index.html

Normal or neutral?

Some sources refer to the normal phase as the *neutral* phase. The two terms may be used interchangeably.

▶ Under the **Expert Assessment** heading, click the **Current Diagnostic Discussion** link.

This page contains the results of ongoing discussions among experts in the field of meteorology who study the El Niño Southern Oscillation (ENSO). Their language can be technical as they discuss different observations. Study the report and see if you can determine whether the experts are predicting the development of El Niño, La Niña, or normal phases.

6. Based on the expert assessment, are normal, El Niño, or La Niña phases predicted for the near future?

7. Based on the expert assessment, what does the near future hold for your chosen area, in terms of precipitation and temperature?

Unit 4
Marine Productivity

In this unit, you will...

- *Discover patterns in global primary productivity*
- *Compare terrestrial and marine productivity*
- *Explore the key resources required for productivity*
- *Correlate variations in marine productivity with limiting resources*
- *Investigate sources of marine nutrients*
- *Synthesize observations to evaluate the causes of dead zones*

The ocean provides up to 20% of the world's food supply.

Bounty from the sea

Activity 4.1

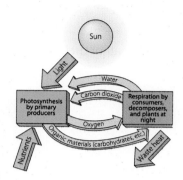

Figure 1. Schematic diagram of photosynthesis.

autotroph—("self-feeder") organism that makes its own food rather than consuming other organisms.

heterotroph—("other-feeder") organism that consumes other organisms.

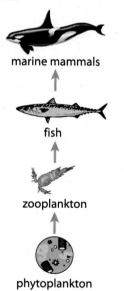

Figure 2. Simple marine food chain. Arrows represent the transfer of energy from one organism to another through consumption.

Seafood makes up 20% of the world's food supply, with over one billion people depending on its resources for survival. As seafood harvests have increased over the past two centuries, populations of some species of marine life have decreased and even become extinct. Given the ocean's vast area, it is difficult to locate, monitor, and track changes in stocks of commercially important fish and shellfish. Thus, scientists frequently use satellites to indirectly assess the health of fisheries and the ocean ecosystem.

A key indicator of the ocean's health is *primary productivity,* the amount of new organic material produced through *photosynthesis* (Figure 1). Photosynthesis is the process by which plant cells containing the green pigment *chlorophyll* use sunlight to convert water and carbon dioxide into the food (sugars and starches) and oxygen needed by most other organisms. Satellites can measure the amount of chlorophyll contained in single-celled plants in the ocean's surface layer, from which we can estimate primary productivity.

Food chains (Figure 2) and more complex food webs (Figure 3) illustrate feeding relationships between organisms in biological communities. At the base of food chains and webs are *autotrophs*, which produce their own food for growth and reproduction through photosynthesis. In the ocean, the primary autotrophs are phytoplankton, microscopic single-celled plants that drift near the ocean surface. The remaining organisms in food webs are *heterotrophs*, which obtain food by feeding on other organisms.

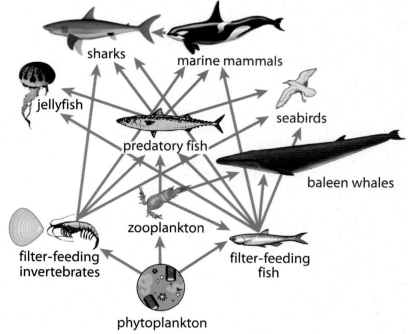

Figure 3. A marine food web.

biotic community— group of interdependent organisms inhabiting the same region and interacting with each other.

anchovies—small fish, similar to sardines, that eat zooplankton.

tuna—large predatory fish that eat other fish.

The preservation of each link in a food web is critical for maintaining diverse and healthy *biotic communities*. However, certain organisms are more critical than others.

1. Examine the complex marine food web in Figure 3. Add anchovies (**A**), tuna (**T**), and humans (**H**) to the figure where you think they fit best. Draw arrows as needed to show consumption of and by other organisms.

2. What do you think would happen if you removed all of the autotrophs from the marine food web?

3. What do you think would happen if you removed one of the heterotrophs such as the predatory fish?

4. How do humans influence food webs?

respiration—process by which organisms oxidize or "burn" food, producing water and carbon dioxide.

Photosynthesis requires four key ingredients: water, sunlight, nutrients, and carbon dioxide. For marine autotrophs like phytoplankton, there is plenty of water available in the oceans. Carbon dioxide is also abundant in ocean waters. It is released as a by-product of *respiration*, and it is readily absorbed from the atmosphere into the ocean. Thus, water and carbon dioxide are not limiting resources for photosynthesis or primary productivity in the ocean. However, marine productivity is controlled or limited by the availability of the other two necessary resources—sunlight and nutrients.

What are nutrients?

Nutrients are chemical compounds that are used by bacteria and plants as the building blocks for organic material. Common nutrients include:

5. How do you think the availability of limiting resources might vary in the ocean (where or when would they be high or low)?

- phosphates (PO_4^-)
- nitrates (NO_3^-)
- silica (SiO_4^-)
- iron (Fe^{3+})

6. Based on the availability of limiting resources, where would you expect phytoplankton to be most productive and least productive? (Near the equator or near the poles? Near the coast or in the open ocean?)

7. How is the productivity of autotrophs important to society? Describe three different ways in which a severe decrease in primary productivity would affect society.

   a.

   b.

   c.

Activity 4.2

Photosynthesis

The general chemical equation for photosynthesis is:

$$6CO_2 + 6H_2O \xrightarrow{\text{sunlight}} C_6H_{12}O_6 + 6O_2$$

carbon dioxide   water   glucose (carbohydrate)   oxygen

primary productivity—the rate at which new organic material is formed by photosynthesis.

NASA

Figure 1. The MODIS (MODerate-resolution Imaging Spectrometer) instrument on the Terra satellite is the latest tool for measuring primary productivity from space.

The life-giving ocean

Phytoplankton are tiny—hundreds of them could fit side-by-side across the width of a human hair—but collectively, they pack a wallop. Phytoplankton are *primary producers*, serving as the first link in almost every food chain in the ocean. They transform water and carbon dioxide into carbohydrates, which they use for producing energy for growth and reproduction. Phytoplankton, in turn, are food for other organisms, passing carbohydrates and other nutrients up the food chain. Because phytoplankton release oxygen during photosynthesis, they also play a significant role in maintaining the proper balance of Earth's atmospheric gases. Phytoplankton produce about half of the world's oxygen and, in doing so, remove large amounts of carbon dioxide from the atmosphere.

Given the central role of phytoplankton in stabilizing the mixture of gases in Earth's atmosphere and in providing food for other organisms, it is important to monitor their location and rate of productivity. Although it is impossible to directly measure the productivity of phytoplankton on a global scale, there are several ways of making indirect calculations. One method relies on the distinctive way that chlorophyll, the green pigment in phytoplankton and other plants, reflects sunlight. By using satellites to measure the chlorophyll concentration of the ocean surface layer, scientists can estimate the rate at which phytoplankton produce carbohydrates (Figure 1). Because carbon is the key element in the process, productivity is measured in terms of kilograms of carbon converted per square meter (kgC/m^2) per year.

Global primary productivity

In this exercise you will examine global *primary productivity* and its relation to the factors that support phytoplankton growth.

▸ Open ArcView, then locate and open the **etoe_unit_4.apr** file.

▸ Open the **Primary Productivity** view.

The **Terrestrial** and **Marine Productivity (kgC/m^2)** themes display the average annual primary productivity for terrestrial and marine environments. During their respective winters, regions near the north and south poles receive little or no sunlight, making satellite measurements impossible. Nonetheless it is reasonable to assume that, with no sunlight, winter primary productivity in these regions is essentially zero.

1. What colors represent areas of highest and lowest productivity?

   a. highest

   b. lowest

2. On Map 1, circle the land areas with the highest productivity using a solid line and the land areas with the lowest productivity using a dashed line.

Map 1—Areas of highest and lowest productivity

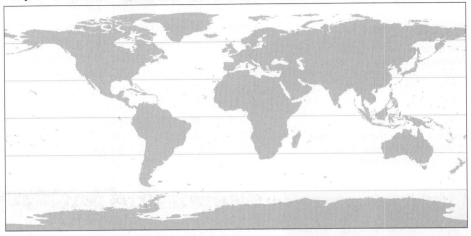

To turn a theme on or off, click its checkbox in the Table of Contents.

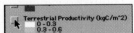

▶ Turn off the **Terrestrial Productivity (kgC/m^2)** theme.

▶ Using the Zoom In tool ⊕, examine, in detail, the areas with the highest marine productivity. When you are finished, click the Full Extent button 🌐 to zoom back out to show the entire map.

3. Mark the areas with highest marine productivity on Map 1 using the label **H** (high), and the areas of lowest marine productivity using the label **L** (low).

4. Where is marine productivity generally:

   a. highest?

   b. lowest?

5. Compare the locations of regions of high terrestrial and marine productivity on Map 1. Describe any geographic patterns or similarities in their distribution.

Productivity and distance from coast

Next, you will examine productivity near the coastline in greater detail.

▶ Turn off the **Marine Productivity (kgC/m^2)** theme and turn on the **Coastal Productivity (kgC/m^2)** theme.

▶ Click the Summarize button 🔲.

▶ In the Summary Table Definition window, summarize the **Coastal Productivity (kgC/m^2)** theme based on the **Distance** field. Include in the table the **Total Productivity (kgC)** field summarized by **Sum** and click the **Add** button. Also, include the **Area (m^2)** field summarized by **Sum** and click the **Add** button again. Select the **All Values** option, then click **OK**. (Note: This may take a while!)

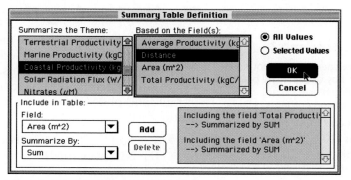

In the resulting summary table, the **Total Productivity (kgC)_Sum** field gives the total productivity and the **Area (m^2)_Sum** field gives the total area for each coastal zone.

6. Record the total productivity and total area for each coastal zone in Table 1. Round productivity and area to the nearest 0.1 trillion. The order of the distance zones in the summary table is not the same as in Table 1, so be sure to record the values in the correct rows. (See note at left for help with converting values to trillions.)

Converting to trillions

Move the decimal 12 places to the left, then round to the appropriate decimal place.

Example:

27656585301787.3

= 27.7 trillion

Calculating average productivity

To calculate average productivity, divide the Total Productivity by the Total Area. Be sure to write both measurements in trillions, so the trillions will cancel each other.

Example:

$$\frac{\text{Total Productivity}}{\text{Total Area}} = \frac{27.7 \text{ trillion kgC}}{211.0 \text{ trillion m}^2}$$

$$= 0.13 \text{ kgC/m}^2$$

Table 1—Marine productivity with distance from coastline

Coastal Zone	Total Productivity (trillion kgC)	Total Area (trillion m²)	Average Productivity (kgC/ m²)
Near (0-320 km)			
Mid (320-640 km)			
Far (640-960 km)			
Open Ocean (> 960 km)	27.7	211.0	0.13

7. Calculate the average marine productivity for each zone and record your results in Table 1. (See left) Round to the nearest 0.01 kgC/m².

8. What happens to the level of marine productivity as the distance from the coastline increases?

Photosynthesis requires sunlight, water, carbon dioxide, and nutrients. If these resources are present in the proper quantities, productivity will be high; if not, productivity will be reduced.

9. Which resource necessary for photosynthesis do you think is most likely to change with distance from the coast to produce the pattern you observe in Table 1? Explain your answer.

▸ Close the Summary Table window.

Understanding the patterns

What are nutrients?

Nutrients are chemical compounds that are used by bacteria and plants as the building blocks for organic material. Common nutrients include:

- phosphates (PO_4^-)
- nitrates (NO_3^-)
- silica (SiO_4^-)
- iron (Fe^{3+})

Micromoles

Micromolarity is a measure of concentration used to describe very weak solutions. A **mole** is 6.02×10^{23} molecules (or atoms), so a micromole is a millionth of a mole, or 6.02×10^{17} molecules. That seems like a lot, but when dissolved in a liter of water (55.5 moles, or about 3.34×10^{25} water molecules) it's only one molecule of nutrient for every 200 million water molecules. That's a weak solution!

Latitude bands

Low latitudes

    0-30° N and S
    near equator

Middle latitudes

    30-60° N and S
    between equator and poles

High latitudes

    60-90° N and S
    near poles

Water and carbon dioxide are never limiting resources in the ocean environment. Thus, marine productivity is governed by the availability of sunlight and nutrients. In this section, you will determine when and where sunlight and nutrients control marine productivity. You will examine nitrates and phosphates, but other nutrients like silica and iron affect productivity in similar ways.

▸ Turn off the **Coastal Productivity (kgC/m^2)** theme and turn on the **Nitrates (µM)** theme.

Nitrates and phosphates are important nutrients that are used by autotrophs for building complex molecules needed for growth and development. The **Nitrates (µM)** theme displays the average annual level of nitrates in the world's oceans in terms of *micromolarity*, or millionths of a mole of nitrate per liter of seawater. (See left.)

10. Which latitude bands have the highest and lowest concentrations of nitrates? (See *Latitude bands* at left.)

    a. Highest

    b. Lowest

▸ Turn off the **Nitrates (µM)** theme and turn on the **Phosphates (µM)** theme.

This theme shows the average annual concentration of phosphates in the world's oceans.

11. Which latitude bands have the highest and lowest concentration of phosphates?

    a. Highest

    b. Lowest

12. Are the patterns for both nutrients similar? If not, how do they differ?

▶ Turn off the **Phosphates (µM)** theme and turn on the **Solar Radiation Flux (W/m^2)** theme.

This theme shows the average annual solar radiation that strikes Earth's surface, in watts per square meter (W/m^2).

▶ Click the Media Viewer button 📽 and open the **Solar Flux Movie**.

This animation shows changes in solar radiation throughout the year. The time of year and the legend appear at the bottom of the image.

▶ View the movie several times.

13. Where is the average solar radiation:
    a. Highest?

    b. Lowest?

14. How does the pattern of nutrient concentration you noted in questions 10 and 11 compare to the pattern of solar radiation? Explain your answer.

▶ Quit the QuickTime Player application, then click the Media Viewer button 📽 and open the **Productivity Movie**.

This animation shows marine productivity throughout the year. The time of year is indicated at the top of the image and the legend appears at the bottom. Black areas at high latitudes are where there was no reflected sunlight during the winter—you can assume these areas have zero productivity.

▶ Study the movie to examine how productivity changes throughout the year. Focus on only the extreme high and low levels of productivity. You will need to play the movie several times to answer the questions below. It may help to focus on answering a single part of each question each time you run the movie.

15. In Table 2, enter the months and season when productivity is highest and lowest in each hemisphere. (See **Earth's seasons** at left.)

Seasons

Earth's seasons are:

• caused by the tilt of Earth's axis.
• opposite in the northern and southern hemispheres.

Dates (typical)	Hemisphere	
	Northern	**Southern**
Dec 21–Mar 19	Winter	Summer
Mar 19–Jun 20	Spring	Fall
Jun 21–Sep 21	Summer	Winter
Sep 22–Dec 20	Fall	Spring

Table 2—Productivity extremes by hemisphere

Hemisphere	Months of high productivity	Season	Months of low productivity	Season
Northern				
Southern				

▶ Quit the QuickTime Player application.

16. Discuss the influence that each of the following factors has on productivity at *high* latitudes, and how productivity in that region may change with the seasons.

   a. Sunlight

   b. Nutrients

17. Discuss how each of the following factors contributes to the productivity pattern at *low* latitudes, and how productivity in that region may change with the seasons.

   a. Sunlight

   b. Nutrients

▶ Turn off the **Solar Radiation Flux (W/m^2)** theme.

In the next activity, you will learn more about marine productivity, the sources of marine nutrients, and the processes that bring nutrients to the surface near the coastlines and in the open ocean.

▶ Choose **File > Exit** to quit ArcView GIS. Do not save your changes to the project file.

Activity 4.3

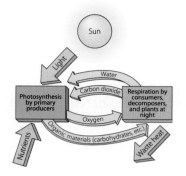

Figure 1. Photosynthesis and respiration.

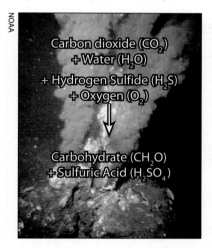

Carbon dioxide (CO_2)
+ Water (H_2O)
+ Hydrogen Sulfide (H_2S)
+ Oxygen (O_2)

Carbohydrate (CH_2O)
+ Sulfuric Acid (H_2SO_4)

Figure 2. Process of chemosynthesis.

Figure 3. Tube worms at hydrothermal vents consume chemosynthetic bacteria.

Resources for productivity

Mean, green food-making machines

With few exceptions, life on Earth depends on *photosynthesis*, the biological process that converts solar energy and inorganic compounds into food. Only autotrophs or "producers," including green plants and phytoplankton, are capable of photosynthesis. They contain the pigment *chlorophyll*, which uses solar energy to convert carbon dioxide and water into carbohydrates (Figure 1). The amount of carbon converted to food by autotrophs is referred to as *primary productivity*. Autotrophs use some of this food immediately, and store the remainder for later use, converting it back to energy through the process of *respiration*. Phytoplankton are consumed by other organisms which are, in turn, consumed by other organisms on up the food chain. Collectively, these consumer organisms are called *heterotrophs*.

Exceptions to the rule

Green plants and phytoplankton are not the only organisms capable of synthesizing their own food. In the last 40 years, scientists have discovered biological communities that are not based on the sun's energy. The autotrophs in these communities are microbes that convert carbon dioxide and water into food using chemicals rather than the sunlight. This process is called *chemosynthesis* (Figure 2). Chemosynthetic bacteria are able to thrive in the high temperatures and pressures of environments like deep sea volcanic vents. Instead of sunlight, they synthesize food using the chemical energy of hydrogen sulfide emerging from the vents. Chemosynthesis supports a diverse community of organisms (Figure 3).

Food for thought—bottoms up!

Food webs illustrate the feeding relationships between organisms in a biotic community. The arrows represent the transfer of energy from one organism to another through consumption (Figure 4). Autotrophs produce most of Earth's atmospheric oxygen as a by-product of photosynthesis, and are the foundation of nearly all terrestrial and aquatic food chains. Thus, primary productivity in terrestrial and marine environments is a key indicator of the overall health of the environment. Our ability to monitor primary productivity has profound economic and environmental importance.

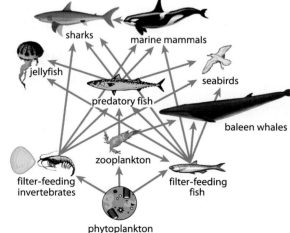

Figure 4. Marine food web.

1. Name the two processes that autotrophs use to synthesize food.

2. What does the number of arrows leading *from* one organism to others suggest about that organism's importance in the food web?

3. What does the number of arrows leading *to* an organism suggest about that organism's likelihood of becoming endangered or extinct?

Resources for photosynthesis

Primary productivity levels vary with season and geographic location. You have identified and examined regions of extremely high and low productivity on land and in the ocean. These patterns of productivity are dictated by the level of the resources necessary for photosynthesis. Autotrophs require carbon dioxide, water, sunlight, and nutrients to photosynthesize (Figure 1). When these resources are not present in adequate amounts for photosynthesis, they are referred to as limiting factors. Below, you will examine the likelihood of each resource to be a limiting factor in productivity.

Carbon dioxide

Carbon dioxide is not likely to be a limiting factor in terrestrial or marine photosynthesis because it is plentiful in the atmosphere, as a result of respiration and human activities. As organisms convert food into energy, they release carbon dioxide (Figure 1). When humans burn fossil fuels like petroleum, coal, and natural gas, huge quantities of carbon dioxide are released into the atmosphere. Carbon dioxide is readily absorbed into the ocean from the atmosphere, so it is in good supply there as well.

Water

Water can often be a limiting factor for productivity in terrestrial environments. Productivity in deserts is typically very low because deserts lack the water needed to sustain all but the hardiest plants. In the ocean, obviously, water is never a limiting factor.

Sunlight

In the context of productivity, the sun's most important role is providing the light energy that drives photosynthesis. The availability of light to autotrophs varies, depending on latitude, season, and time of day. Therefore, there are places and times where sunlight is a limiting factor.

Nutrients

Autotrophs use inorganic compounds containing nitrogen, phosphorus, potassium, calcium, silicon, and iron to build organic molecules. The concentrations of these *nutrients* vary throughout the environment. Nitrogen and phosphorus are particularly important because they are required in large quantities and play a critical role in growth and reproduction. Inorganic forms of nitrogen are the building blocks for amino acids, proteins, and genetic material (DNA, RNA). Similarly, phosphorus is an essential component of energy transport molecules (ATP), genetic material, and structural materials (bone, teeth, shell).

Organic or inorganic?

The terms *organic* and *inorganic* originally came from the idea that chemical compounds could be divided into two categories: those coming from or composed of plants or animals (organic) and those extracted from minerals and ores (inorganic).

Chemists now think of organic compounds as those that contain carbon. This definition works well for most compounds, but there are exceptions. For example, carbon dioxide is a carbon-based compound, but it is not considered organic.

Despite the abundance of nitrogen gas in the atmosphere, it is often a limiting resource because it can be utilized by autotrophs only in certain forms. In the *nitrogen cycle*, nitrogen gas is converted to its most useful form, nitrates, through a series of reactions carried out by bacteria and fungi in the soil (Figure 5). Nitrates may be utilized by land plants, or may be carried to the ocean in ground water or runoff and used by phytoplankton.

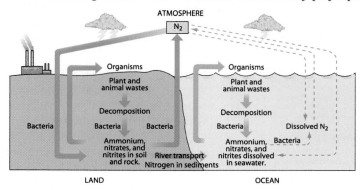

Figure 5. The Nitrogen Cycle.

Unlike nitrogen, phosphorus never exists as a gas. Phosphorus originates in rocks in the Earth's crust in the form of phosphate salts, which are liberated from rocks by weathering (Figure 6). Phosphates are also an ingredient in some detergents and fertilizers. Precipitation carries phosphates into the soil, and runoff and ground water transport them to the ocean.

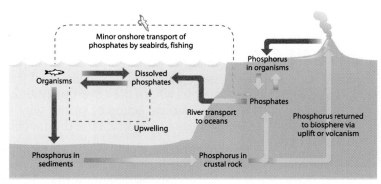

Figure 6. The Phosphorus Cycle.

4. Various human activities contribute additional nutrients to the ocean ecosystem. How might these additions influence marine productivity?

Sources of nutrients

Most of the nutrients utilized by phytoplankton in surface waters originate on land. Nitrates and phosphates in the soil dissolve easily in water. These nutrients are easily leached or removed from the soil by precipitation and runoff, and are carried by ground water, streams, and rivers to the ocean. Near the coasts, phytoplankton thrive on the nutrients entering the ocean from land, resulting in high productivity. However, large portions of the nutrients entering the ocean are not utilized and eventually sink to the ocean floor, accumulating in the sediment.

Not all nutrients enter the ocean from land. When marine plants and animals die they sink to the bottom, where decomposition liberates the nutrients, making them available for use again. However, except in shallow waters over the continental shelf, the nutrients on the ocean bottom can be utilized only when they are brought from the depths to the surface via *upwelling*, the upward movement of deep, cold bottom water to the surface. Coastal upwelling occurs in nearshore environments where strong winds blow parallel to the shore (Figure 7). Ekman transport causes the surface currents to deflect away from the shore, which pulls deep, cold, nutrient-rich water toward the surface.

Upwelling also occurs in the open ocean, near the equator. Equatorial upwelling is controlled by ocean currents—water moving westward on either side of the equator is deflected toward the poles and is replaced by cold, nutrient-rich water from below (Figure 8).

In the previous examples, upwelling is characterized by either winds or currents that move surface water and allow deeper water to "well up" in its place. The action of the wind or current is especially important in regions where there is a strong *thermocline*, a region where the water temperature changes rapidly with depth (Figure 9). Waters near the equator typically have a strong thermocline that traps nutrients in the cold, deep waters below, where they eventually sink to the ocean floor. At low latitudes, the mechanical action of strong winds or currents is necessary for nutrient-rich waters to penetrate the thermocline and rise to the surface.

In contrast, in high-latitude regions, upwelling can occur without the help of strong winds. These regions receive little or no sunlight, resulting in a weak or nonexistent thermocline. As dense water sinks, it is replaced by an equal volume of nutrient-rich water rising from the adjacent depths.

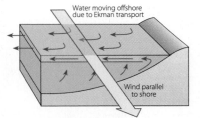

Figure 7. Coastal upwelling.

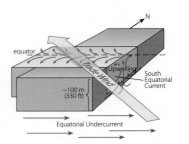

Figure 8. Equatorial upwelling.

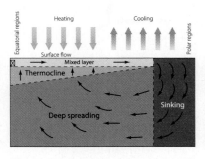

Figure 9. Schematic cross-section of the ocean from equatorial to polar regions, showing change of depth to thermocline with changing latitude.

5. What process makes nutrients in deep ocean sediments accessible to marine phytoplankton drifting near the ocean surface?

6. Explain how the strength of the thermocline affects the availability of nutrients to primary producers.

7. Human activities can increase levels of nutrients in the ocean. Describe how human activities could influence the levels of *other resources* important in photosynthesis.

Too much of a good thing?

Previously, you observed that a shortage of one or more resources can decrease productivity. It may surprise you to learn that there can also be too much of a good thing when it comes to photosynthesis. For example, if the amount of solar energy received by the ocean were to intensify dramatically, tropical oceans could become too warm for photosynthesis to proceed properly.

Similarly, nutrients can be present in quantities so high that they become harmful. High levels of nutrients in coastal waters may cause a sharp increase in phytoplankton that reduces light penetration, clogs the gills of marine organisms, and pollutes the water with waste products, some of which are toxic to other marine life. This may lead to *hypoxia*, a dramatic decrease in dissolved oxygen, or even *anoxia*, a total absence of dissolved oxygen, which can drive away mobile organisms and kill *sedentary organisms*.

sedentary organisms—organisms that are attached to a surface and cannot move freely.

Two conditions contribute to hypoxia in subsurface waters.

- **Stratification** or layering of the water column. Nutrient-laden runoff flowing into the ocean is less dense than salt water and floats on the surface. In summer, warm weather and calm seas inhibit mixing of the shallow, warm water with deeper, colder water. As a result, the oxygen from photosynthesis remains at the surface.

- **Increased decomposition** at depth, due to higher surface productivity. Phytoplankton die or are eaten by other organisms, creating large amounts of organic waste that sink to the ocean floor. As the waste decomposes, nutrients are recycled but oxygen is also consumed, creating hypoxic conditions.

Figure 10. Location and extent of the 1993 Mississippi River dead zone.

estuary—body of water where a river meets the ocean, mixing fresh river water with ocean water.

Is the Mississippi River dead zone unique?

In the United States alone, more than half of the estuaries experience hypoxia during the summer; up to a third experience anoxia.

The Gulf of Mexico is an important US commercial fishery. In 2002, the Gulf accounted for 16.9% by weight (1.7 billion lbs.) and 24.7% by dollar value ($693 million) of the entire US fish and shellfish catch. (National Marine Fisheries Service, NOAA) Each summer, a large hypoxic region known as the *Mississippi River dead zone* forms off the coast of Louisiana (Figure 10). The lack of dissolved oxygen in this region causes both the quantity and the diversity of economically-important marine life to decrease dramatically, with potentially dire economic consequences.

8. Describe how an increase in nutrient levels can actually *lower* marine productivity in the marine ecosystem.

9. Name two human activities that could result in an overabundance of nutrients being delivered to coastal waters. Explain your answer.

10. How might reduced light penetration affect primary producers as well as other animals higher up the food chain.

In the next activity, you will examine changes in the Mississippi River dead zone from 1986 to 1993, and investigate patterns in the distribution of dead zones around the world.

Activity 4.4

Dead zones

There are parts of the ocean where no fish swim, and where the bottom may be littered with the remains of bottom-dwelling crabs, clams, and worms. These areas, known as *dead zones*, pose a growing environmental concern. The causes of dead zones are complex and involve pollution in the form of excess nutrients from agriculture, industry, urbanization, and other sources.

Creating a dead zone

One of the largest dead zones in the world occurs in the northern Gulf of Mexico, where the Mississippi River flows into the open ocean off the coast of Louisiana. Starting in the 1950s, the waters of the Mississippi River began to show signs of elevated nutrient levels, particularly nitrogen compounds from agricultural runoff in the Mississippi River basin. At the same time, a dead zone began to form adjacent to the Mississippi delta in the Gulf of Mexico.

In this activity, you will look at the size of the Mississippi River dead zone over time, and compare it to dead zones on the east coast of the United States and around the world. This will help you to better understand how and where dead zones form, and to predict where they may develop in the future.

▶ Open ArcView GIS, then locate and open the **etoe_unit_4.apr** file.

▶ Open the **Mississippi River Dead Zone** view.

To turn a theme on or off, click its checkbox in the Table of Contents.

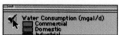

watershed—an area of land that drains downslope toward the lowest point.

The Mississippi River dead zone

This view shows the Mississippi River and the Mississippi River *watershed*. The 31 states that are either partially or completely within the Mississippi River watershed boundary are shown in gray.

The Mississippi River watershed covers more than half of the 48 contiguous states, and many activities affect the quality of the water before it empties into the Gulf of Mexico. Next, you will examine how water is used, to identify possible sources of the excess nutrients in the Mississippi River dead zone.

▶ Turn on the **Water Consumption** theme.

This theme consists of pie charts for the 48 contiguous states, each showing the percentage of water consumed by six major water use sectors. (See left.) The overall size of each pie represents the total amount of water consumed by that state, and the slices represent the percentage of water used by each water use sector.

Water use sectors
- **Commercial**—facilities and institutions including hotels, restaurants, hospitals and schools.
- **Domestic**—household use.
- **Industrial**—producing steel, chemicals, paper, plastics, minerals, petroleum, and other products.
- **Power**—steam-driven electric generators. (Does not include hydroelectric power.)
- **Mining**—extracting minerals, oil, and natural gas.
- **Agriculture**—raising animals and irrigating crops.

1. Which activities are the primary consumers of water in states within the Mississippi River watershed?

    a. Eastern portion of the watershed.

    b. Western portion of the watershed.

2. How might these activities contribute to nutrient enrichment of the Mississippi River?

▶ Turn off the **Water Consumption** theme and turn on the **Precipitation (cm/yr)** theme.

This theme shows average precipitation, in centimeters per year. Areas with low precipitation are shown in yellow and orange, and areas with high precipitation in green and blue.

3. Where is precipitation in the Mississippi River watershed:

    a. Highest?

    b. Lowest?

Runoff from the land is one mechanism for transporting nutrients from farms and fields to the river.

4. Based on the precipitation patterns, predict where you would expect runoff to be highest.

▶ Turn off the **Precipitation (cm/yr)** theme and turn on the **Runoff (cm/yr)** theme.

This theme shows average runoff within the Mississippi River basin, in centimeters per year. Dark blue represents high runoff and light blue represents low runoff.

5. Which parts of the Mississippi River basin have the highest annual runoff?

6. Describe any relationships you observe between the precipitation and the runoff patterns.

7. How might levels of precipitation and runoff influence the size of the Mississippi River dead zone?

▶ Turn off all themes except the **Mississippi Watershed** and **States** themes.

Development of the Mississippi River dead zone

In this section, you will look at the extent of the dead zone at four different time periods, and compare the size of the dead zone to the climate data for the Mississippi River watershed for that same time.

▶ Click the QuickLoad button 🔍 and load the **Gulf of Mexico** extent.

In 1985, scientists began sampling the water off the Louisiana coast to monitor the extent and characteristics of the Mississippi River dead zone.

To activate a theme, click on its name in the Table of Contents.

▶ Turn on and activate the **1986 Dead Zone** theme.

▶ Using the Identify tool 🛈, click inside the dead zone boundary to display the area of the dead zone for that year.

8. Record the area of the dead zone in Table 1. Round to the nearest 1 km².

Table 1—Mississippi River dead zone data

Year	Area (km²)	Area Rank	Drought Appearance
1986			
1988			
1990			
1993			

▶ Close the Identify Results window and turn off the **1986 Dead Zone** theme.

▶ Repeat these steps to record the areas of the dead zones for 1988, 1990, and 1993 in Table 1. See Figure 1 for the location of the 1988 dead zone.

9. In Table 1, rank the areas of the dead zones from 1 (largest) to 4 (smallest).

Figure 1. Location of 1988 Mississippi River Dead Zone.

> ▶ Click the QuickLoad button 🔍 and load the **US 48** extent.
>
> ▶ Turn off all of the **Dead Zone** themes and turn on the **1986 Drought Index** theme.

In the **Drought Index** themes, green (index > 0) represents areas that are wetter than normal, and brown (index < 0) represents areas that are drier than normal.

10. In Table 1, describe the overall climate of the Mississippi River watershed for each of the four years using one of the following descriptions: *very wet, wet, dry,* or *very dry*. Turn the **Drought Index** themes on and off as needed to complete the table.

11. Based on Table 1, how do the areas of the dead zones appear to respond to the climate conditions in the Mississippi River watershed?

12. Explain the changes in the area of the dead zone in terms of runoff and nutrients in the Gulf of Mexico.

The Mississippi River dead zone is not the only one in the U.S. or the world. Next, you will examine potential factors that may contribute to the formation of other dead zones.

Global dead zones

> ▶ Click the QuickLoad button 🔍 and load the **Global Dead Zones** view.

This view shows the locations of dead zones throughout the world.

13. Describe the locations of the two primary clusters of dead zones.

Dead zones and population

Some scientists have noted that dead zones are related to the human populations of the adjacent continents. Many believe that dead zones may become the most significant human impact on oceans and ocean ecosystems in the 21st century. In this section, you will investigate this idea and the ways in which increasing population influences the size and number of dead zones.

14. If dead zones are related to areas with large populations, where on the map would you expect to find areas with large populations?

You can test your prediction by looking at the global population distribution, and see how it relates to the current distribution of dead zones.

▶ Turn on the **Global Population (people/sq degree)** theme.

The **Global Population (people/sq degree)** theme shows the population for 1°× 1° *cells* (approximately rectangular regions within a grid) over Earth's surface.

How large an area is a 1°x 1° cell?

The spacing between latitude lines is fairly constant, but the spacing between longitude lines decreases as you move away from the equator. Thus, the areas of 1° x 1° cells get smaller with increasing distance from the equator.

At the equator, a 1° x 1° cell has an area of about 12,500 km². At 60° latitude, the area of a 1° x 1° cell is only half that size, about 6,200 km².

15. Some scientists argue that high population causes dead zones. Examine the global patterns of population density and dead zones. Describe any observations that:

a. *Support* (provide evidence in favor of) the idea that high population causes dead zones.

b. *Refute* (provide evidence against) the idea that high population causes dead zones.

There appears to be a connection between large populations and the formation of dead zones, but it is not a clear or consistent relationship. One possibility is that there is some important difference in the populations, other than their sizes, between the regions where dead zones form and where they do not form. Next, you will explore the possibility that *economic differences* between populations are a factor in the formation of dead zones.

▶ Turn off the **Global Population (people/sq degree)** theme and turn on the **Gross Domestic Product (billion $US)** theme.

Gross Domestic Product (GDP) is the total value of goods and services produced by a nation within that nation's boundaries. Indirectly, it is a measure of the resources (energy, raw materials, people) available to and utilized by that nation.

Now you will determine the average GDP of countries adjacent to dead zones and compare it to the average GDP of countries without dead zones.

Select By Theme button

The Select By Theme button is located on the Button Bar (the top row of buttons).

▶ Click the Select By Theme button 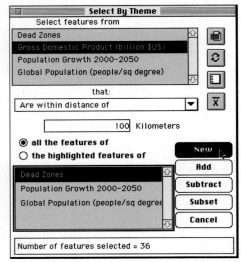 (location shown at left) and select features from the **Gross Domestic Product (billion $US)** theme that **Are within distance of 100 km** of **all the features of** the **Dead Zone** theme.

▶ Click the **New** button but do not close the Select By Theme window.

All of the countries with dead zones are highlighted in the view.

▶ Click the Statistics button ⊠ in the Select By Theme window and calculate statistics for **the selected features of** the **Gross Domestic Product (billion $US)** theme using the **2002 GDP (billion $US)** and **Population** fields. (Hold down the shift key to select multiple fields.) Select the **Basic** output option and click **OK**.

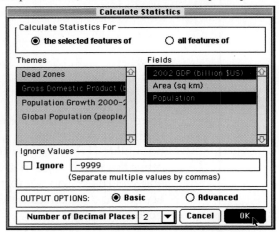

In the statistics window, the number of countries adjacent to dead zones is given as the **Number of Features** and the total GDP and population of those countries is given as the **Total**.

16. Record the number of countries (**Number of Features**), the GDP (**Total**), and the Population (**Total**) in Table 2. Record the values in the units shown in the table, to the nearest whole number for Total GDP and to the nearest 0.01 billion for Total Population.

Table 2—Relationship of per capita GDP and presence of dead zones

	Number of countries	Total GDP (billion $US)	Total Population (billions)	Per Capita GDP
Near dead zones				
Not near dead zones				

▶ Close the Statistics window but do not close the Select By Theme window.

▶ Click the Switch Selection button ⟳ in the Select By Theme window.

The countries that are *not* near dead zones will be highlighted.

▶ Click the Statistics button x̄ in the Select By Theme window again and repeat the previous statistics calculation.

17. Record the number of countries, GDP, and population values for countries not near dead zones in Table 2.

▶ Close the Statistics and Select By Theme windows.

per capita—for each person.

Why use per capita GDP?

A large country may have a higher GDP than a smaller country simply because it has more people, even though the people in the smaller country may individually produce more goods and services. To correct for this, we use per capita GDP—the amount of goods and services produced by a country, divided by the number of people in that country.

18. Calculate the Per Capita GDP for the countries near dead zones and the countries not near dead zones and record them in Table 2. (Divide the Total GDP by the Total Population.)

19. How does the per capita GDP values differ for the two groups of countries?

20. How might GDP be a better indicator of the future development of dead zones than population?

21. In 2000, the world's population was about 6.1 billion. By 2050, it is expected to increase to 8.9 billion, with the largest growth in countries with low per capita GDPs. If the total global GDP—the sum of all the goods and services produced by all nations—increases at a similar rate, what effect do you think it will have on the size and location of dead zones?

▶ Choose **File > Exit** to quit ArcView GIS. Do not save your changes to the project file.

Activity 4.5

Searching for solutions

The Gulf of Mexico is in bad shape, sick from the effects of industrial and agricultural pollution, population growth, and urban development. The Gulf receives 300 trillion gallons of runoff each year that contain a vast collection of pollutants originating from factories, hog waste ponds, heavily-fertilized farms, golf courses, and residential lawns, as well as oily grime from urban runoff.

Although these chemicals pose serious health hazards, the added nutrients pose the biggest problem, triggering a series of events leading to the formation of the Mississippi River dead zone. Excess nutrients have also been implicated in the deaths of coral reefs, decline of sea grass beds, occurrence of red tides, and the declining health of estuaries around the Gulf of Mexico.

estuary—body of water where a river meets the ocean, mixing fresh river water with ocean water.

What are red tides?

Red tides are caused by seasonal reproductive surges, or blooms, of certain species of marine algae. During a bloom, colored pigments in these tiny one-celled plants discolor the ocean surface, giving it a reddish-brown appearance.

Most of these algae species are harmless, but a few produce potent chemical neurotoxins (poisons that affect the nervous system). These toxins cause widespread fish kills, contaminate shellfish, and can be deadly to humans and other animals that eat contaminated seafood.

1. Describe the cascade of events that leads to the formation of a dead zone, beginning with the addition of excess nutrients. Be sure to discuss how the presence of excess nutrients affects primary producers and other consumers in the food web.

Whereas the Mississippi River dead zone is the largest in the United States, it is not the only one. Next you will determine the effect your community has on dead zones off the US coasts.

▶ Open ArcView GIS, then locate and open the **etoe_unit_4.apr** file.

▶ Open the **US Dead Zones** view.

This view shows the major rivers and watersheds in the contiguous United States. Find the approximate location of your city or town on the map.

▶ Activate the **Major Rivers** theme and click on the major river nearest to your town using the Identify tool [⊙]. The Identify Results window will appear, listing information about the river you selected.

2. What river carries runoff from your region to the ocean?

▶ Close the Identify Results window.

▶ Activate the **Major Watersheds** theme and click on the watershed in which your town is located using the Identify tool [⊙]. The Identify Results window will appear, listing information about the watershed you selected.

3. What is the name of the regional watershed in which your town is located?

▶ Close the Identify Results window.

Science can tell us a lot about how dead zones are created and their impact on the environment. However, cleaning up dead zones requires significant changes in our behavior as a society.

Search the Internet to learn more about the sources of pollution, the economic impact of dead zones, and the technological solutions to the problem, then answer the questions below. Some starting web sites are provided on the next page.

4. Discuss how your city or town contributes to the formation of a dead zone. Think about local residential, commercial, agricultural, and industrial practices.

mitigate— to make less severe.

5. Discuss how you might mitigate the negative impact of some of these activities.

Media Viewer shortcuts
To open these web pages from within ArcView, click the Media Viewer button [%] and choose the appropriate entry from the list of web sites.

[%] **Exploring Solutions**

[%] **Dead Zone Research**

[%] **Hypoxia**

[%] **Red Tides**

[%] **Gulf in Peril**

Web sites on the Gulf of Mexico dead zone

Potential Solutions for Gulf of Mexico's "Dead Zone" Explored

Discusses the impact of the dead zone and ecological/technological approaches to reducing the nutrient flow into the ocean.

> **http://researchnews.osu.edu/archive/hypoxia.htm**

Environmental Literacy Council

Provides an overview of the problem and links to major government and research groups investigating the problem and solutions.

> **http://www.enviroliteracy.org/article.php/1128.html**

National Center for Appropriate Technology (NCAT)

Discusses causes of hypoxia and human activities that contribute to it. Also discusses approaches to reducing nutrient flow to the ocean.

> **http://www.ncat.org/nutrients/map.html**

The Gulf of Mexico Dead Zone and Red Tides

Discusses the effect of the dead zone on the quality and quantity of fish and seafood stocks in the Gulf of Mexico.

> **http://www.tulane.edu/~bfleury/envirobio/enviroweb/DeadZone.htm**

Deep Trouble - The Gulf in Peril

Series of articles in the Naples (Florida) Daily News outlining all the issues facing the Gulf of Mexico and potential solutions.

> **http://web.naplesnews.com/deeptrouble/index.html**

▶ Choose **File > Exit** to quit ArcView GIS. Do not save your changes to the project file.